Superintending for Contractors: *How to Bring Jobs in On-time, On-budget*

Paul J. Cook

Illustrations by Carl W. Linde

Copyright 1987

R.S. Means Company, Inc.
Construction Publishers & Consultants
Construction Plaza
63 Smiths Lane
Kingston, MA 02364-0800
(781) 422-5000

The book and cover were designed by Norman R. Forgit. Illustrations by Carle W. Linde.
Cover photograph by Norman R. Forgit.

Printed in the United States of America

20 19 18 17 16 15 14 13

Library of Congress Catalog Number 87-165874

ISBN 0-87629-272-4

Table of Contents

Part III
Superintending
the Trades

Part IV
Appendix

Foreword

A superintending career traditionally began in the field where management skills were learned by direct exposure, acceptance of responsibility, and stairstep promotions. Approximately two thirds of the work on a project was once done by employees on the payroll of the superintendent's own company. The superintendent's job was to supervise their work and to monitor the performance of the subcontractors who carried out the remaining one third of the work. The responsibility for scheduling the job, soliciting the subs, and dealing with contracts, change orders, and purchase orders belonged to the project manager in the office.

The distribution of work has changed, however, so that two thirds of the work is now being subcontracted out, and only one third done by the employees of the superintendent's own company. Today's superintendent has become a field project manager, directing and coordinating a much larger percentage of subcontractors, and overseeing the administration of contracts, change orders, and purchase orders. It is the purpose of this book to examine the complex role of the superintendent/field project manager, and to provide guidelines for the efficient organization of this job. The focus is the superintendent's role as a company person successfully fulfilling and completing the contract within the requirements of cost, time, and quality. The means to this goal are efficient planning, detailing, scheduling, purchasing and production.

Part I of the book is a general introduction to the superintendent—his personality, qualifications and relationships in the project. Part II begins with the superintendent's responsibilities for planning the project before the work begins, and continues with the methods used to manage the project once it is under way. Part III covers some of the superintendent's technical concerns throughout the construction process. The Appendix contains a collection of useful references, including some mathematic formulas required for calculations on the job.

This book follows *Estimating for the General Contractor* and *Bidding for the General Contractor*, but stands independently as a reference for both superintendents and those who interact with the superintendent in the course of a project.

Part I

The Role
of the
Superintendent

Chapter 1
Qualifications

Construction know-how is a basic qualification of the superintendent. Taken alone, however, it might not be enough. An experienced carpenter may possess the construction know-how needed to construct a project, yet still fall short in other important areas. Motivation, leadership, and organizational and decision-making ability are other good qualifications, along with a talent for public relations.

Motivation provides the spirit to accept a challenge, and leadership lies in the natural and comfortable acceptance of the managing role. The many diverse contributors to a project look to the central, on-site, authorized agent of the construction company for information, dates, directions and approvals. The field project manager is their contact.

Decision making routinely includes accepting or rejecting delivered materials, hiring and firing workmen, ordering work activities, and choosing construction methods and equipment. The ability and good judgment to make these kinds of decisions distinguishes the role of the superintendent from that of the crew leaders. Listed below are further examples of decisions that come before the superintendent:

1. Opposing or agreeing to demands or requests from inspectors and sponsor representatives for performances outside the scope of the contract.

2. Enforcing the subcontractors' responsibilities for the performance of work that is delayed, disputed or refused for a variety of reasons (see Chapter 16 for an in-depth discussion of subcontractor management).

3. Ordering a nonperforming subcontractor off the job and requesting a replacement. This step is usually taken only after consulting the main office to be sure all contractual and financial ramifications have been considered. For example, the office may have this particular subcontractor scheduled for another, future project. Thus, a decision to replace him now may affect more than one project.

4. Reviewing and adjusting construction methods that may involve great cost or risks to safety.

Public relations talent is valuable in carrying out these decisions, particularly the unpopular ones. The superintendent is given the opportunity to practice public relations in daily negotiations and communications with sponsors, subcontractors, union officials, and others. Chapters 4 and 5 cover both public relations and the negotiating process in more detail.

The ability to plan and organize is also a basic qualification of the superintendent. Only certain portions of planning can be delegated to assistants; the greater part should be done by the superintendent personally. This aspect of the work typically includes the following activities:

1. Studying the drawings, specs, bid package, and subcontract in detail, becoming intimately acquainted with the project and with previously planned sources of materials and labor. This procedure is covered more fully in Chapter 9—"Introduction to the Project".

2. Drawing up a construction progress schedule. More than anyone else, the superintendent uses this schedule as a tool in starting subcontractors and suppliers as early as possible and directing each of them to the shortest possible completion times. The schedule can also be a public relations asset as it helps to satisfy the sponsor's need to know the date of project completion (see Chapter 14—"Time Controls").

3. Selecting the most suitable location for the construction office, yard, storage sheds, and staging area (see Chapter 6— "Planning the Working Area").

4. Contacting individual subcontractors and suppliers to request submittal information, such as shop drawings, samples, and catalogue data. Drawing up a submittal schedule (see Chapter 14).

5. Creating a manning schedule for company-employed workers (see Chapter 15—"Productivity and the Manning Chart").

6. Scheduling meetings with subcontractors' representatives to prepare them for starting, overlapping, and stopping. With advance warning and cooperation in scheduling, subcontractors can assist each other to mutual advantage (see Chapter 11— "Administrative Meetings").

Chapter 2
Responsibilities

A construction company may be considered as having two main activities; these are: (1) the procurement of contracts and (2) the fulfilling of contracts. The two previous books in this series, *Estimating for the General Contractor* and *Bidding for the General Contractor* were directed to the procurement—the first of the company's functions. This book deals exclusively with the second activity—the fulfilling of contracts as directed by the superintendent.

The simplest of construction companies functions most efficiently as a partnership; one member is in charge of the office (procurement) and the other is in charge of the field (fulfilling). In a medium size or larger company, the partners expand into teams of employees, as shown in Figure 2.1.

Whether the company is set up as a proprietorship, a partnership, or a corporation, the division of responsibility between the procurement of a contract and its fulfillment is constant. In a structurally stable company organization, vacancies in superintending positions (unlike true partners) may be filled by hiring. Likewise, should a surplus of superintendents occur due to a shortage of contracts, it is lessened by laying off. The movement of superintendents within the construction industry is typical of the executive turnover found in all industries.

Figure 2.1

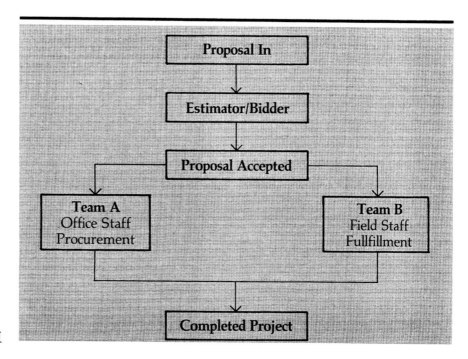

The superintendent's job is to give physical reality to the designing, planning, calculating, bidding, and contracting that have been done in preparation for the job. The superintendent has great autonomy in his project and is the most visible person associated with it. Figure 2.2 shows the heirarchy involved in a construction project organization. Communication takes place between the superintendent and the foremen in one direction and between the superintendent and the sponsor in the other. There is three-way communication between the superintendent, his home office, and the sponsor. There is two-way communication between the superintendent and each of his own foremen as well as each foremen of the various subcontractors.

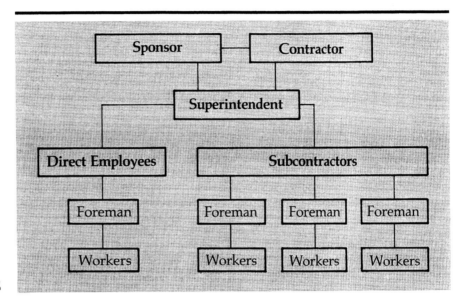

Figure 2.2

The Superintendent as Director

When the superintendent takes on a project, it is the final link in a chain of events that began with the sponsor's idea, and travelled through financing, design, bidding, and contract administration. The project documents (drawings, specifications, and any previously negotiated contractual agreements) received by the superintendent contain many elements and involve several groups and divisions (subcontractors, suppliers, etc.). The job of the superintendent is to direct the flow so that the project proceeds according to the requirements of those documents. Because subcontractors have the required skills as well as a legal and financial commitment, a project theoretically builds itself under the critical directorship of the superintendent.

The Superintendent as Inspector

Contributors to a construction project do more than pass through under the direction of a superintendent. They bring a variety of materials and perform diverse scopes of work. Materials and workmanship must meet certain quality standards acceptable to:
- the sponsor's agent
- the sponsor's architect and engineer
- the sponsor's inspector
- the local building inspector
- state regulatory agents (safety, etc.)
- the subcontractors
- the prime contractor (the superintendent's employer)
- the superintendent

In order to meet the various approvals as they become due, the superintendent continually inspects all materials and workmanship, accepting this, rejecting that, and obtaining compliance from the various contributors by methods which are described in Chapter 19.

Due to conflicts of interest between the various parties, agencies, and contributors, exact quality standards are not easily established. Figure 2.3 depicts the conflict, which is similar to that found in every industry. The buyer (sponsor) demands the most possible from the seller (contractor). The architect and engineer demand the highest possible quality of materials, workmanship, and aesthetics. Regulatory agencies and local building inspectors demand at least minimum compliance with codes. It is the interpretation of this minimum level that is often the subject of debate.

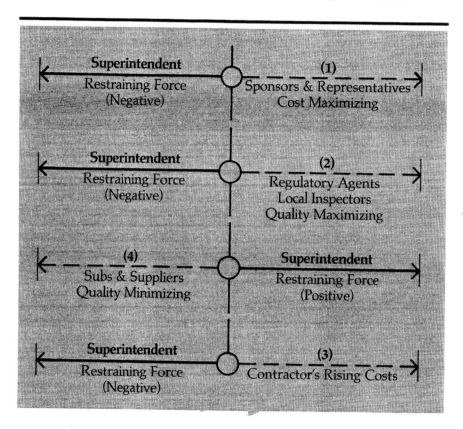

Figure 2.3

Subcontractors and material suppliers are "sellers", primarily to the sponsor (through the superintendent). Figure 2.4 shows the buyer-seller relationship between contributors to a project. A tug-of-war exists between the forces that lean toward maximum and those toward minimum quantity and quality of materials and workmanship. Each party is both buyer and seller; each buyer holds the next lower seller responsible for the performance and quality of all lower tiers of contributing buyers and sellers.

The superintendent exerts pressure on the subcontractors to meet a certain standard of excellence, judged by himself as satisfying the literal requirements of the drawings, specifications and all "reasonable" expectations of the sponsor's architect/engineer. The portions of the project performed by the superintendent's own company must meet the same requirements.

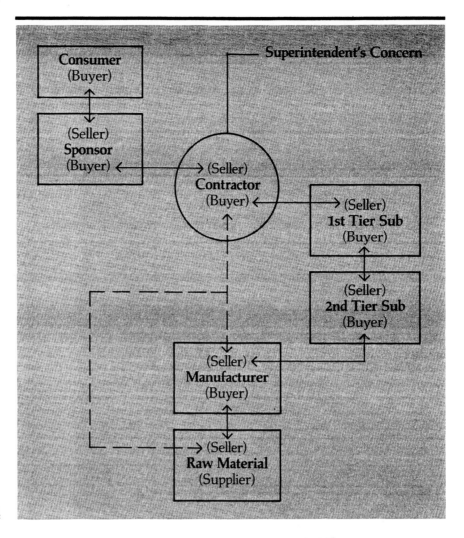

Figure 2.4

Figure 2.3 shows how the superintendent attempts to counteract the upward, excessive demands of the sponsor, while minimizing the bureaucratic red tape in order to reduce the rising costs of his company's own budgeted work. His attitude toward subcontractors is reversed since, as a rule, he must demand higher standards of performance than are automatically supplied. Without strong management by the superintendent, a project will tend to drop in quality and rise in cost.

Errors, Oversights, and Accidents

Errors might involve arithmetic, omission, or faulty judgment. Although every construction professional is susceptible to error, each has a special area of vulnerability. To the estimator, mistakes of arithmetic and omission are an occupational hazard. The bidder's hazard comes in the form of judgment and oversight. The superintendent must watch out for mistakes of judgment in choosing methods, equipment, and safety provisions, and must also avoid slips involving the management of the project's numerous contributors. Following is a list of some of the unfortunate events—and consequences—for which the superintendent may be held responsible.

1. Errors in layout. Extensive construction work may have to be demolished and replaced as a result.

2. Collapses due to faulty formwork, shoring, bracing, connection, or material weaknesses.

3. Worker injuries resulting from inadequate safety observances.

4. Damages from the elements (wind, rain, flooding, heat, or cold), which could have been anticipated and prevented using protection such as covering, bracing, barricading, or sandbagging.

5. Broken underground piping or electrical lines caused by a lack of proper investigation or poor marking of lines set at the beginning of the project.

6. Incorrect materials, quantities, or workmanship—the result of misinterpreted drawings and specifications.

Everyone makes mistakes, but not everyone is prone to them. The person who rises to the position of superintendent has impressed his employer as one who, while not infallible, does have a good working knowledge and control of the projects he manages.

Chapter 3
Ethical Considerations

In the office, during bidding, unethical practice tends to be a product of competition for contracts. It takes a different form in the field, where it is motivated by the pursuit of personal favors and small material benefits. Nearly all profits from unethical practices are short term; long term benefits call for an ethical choice for a future, more meaningful gain. Theoretically, the superintendent always has a choice between the ethical and the unethical, though it may not always be clear. Just as his technical knowledge and organizational skills improve with experience, so does his ability to recognize the unethical, to evaluate the rewards and penalties in both the short and long run, and to choose well.

It is convenient, although an oversimplification, to state that an unethical act involves knowingly and willfully harming one or more persons to the benefit, profit, or advantage of others. To be ethical, a superintendent should be neutral. From that neutral position, he can do someone a good turn with no strings attached and no harm to anyone else. However, to enrich himself or others at someone else's expense may result in both legal and economic penalties. Economic penalties can take the form of:
- Loss of personal or company reputation
- Loss of goodwill which might affect the ability to obtain future contracts

There are, of course, also psychological penalties, such as loss of self esteem and peace of mind.

Assuming that the superintendent is honest, his main areas of vulnerability might be unethical inspectors, gift-giving subcontractors, profit-sharing suppliers who would pass off inferior materials, and non-productive friends on the work force. A more difficult problem is one in which the superintendent is expected, by his employer or company manager, to be actively and agressively unethical. Examples of this kind of activity are:
- Giving gifts to civil service officials or employees.
- Using inferior materials or inferior workmanship.
- Overcharging on change orders.

Witnessing   General construction projects obtained by competitive bidding tend to be committed to tight budgets and completion schedules, and are composed of many subcontracts. This combination of circumstances will often lead to disputes and lawsuits. More often than any other person, the superintendent is a witness in these kinds of events and in any hearings that take place. The superintendent may be called upon as a witness in cases such as:

- Worker injuries
- Trade union matters, such as jurisdictional disputes
- Structural strains or collapses
- Charges and claims made by owners of properties adjacent to the site
- Suits following cancellation of subcontracts for non-performance
- Investigations of noncompliance with spec requirements

Considering the complex network of relationships making up a project, it is surprising how few legal actions involve the superintendent—other than as a witness.

The best approach is to be a good, unemotional witness, presenting the facts briefly and efficiently. When events reach the point of litigation, they are no longer in the superintendent's hands; the best outcome results from honest impartiality.

Chapter 4
Labor Relations

"Cracking the whip" to increase production was a method that began phasing out more than a century ago, but the expression is still used to describe applied pressure (pushing, rather than leading). Through the power of labor organizations and other socio-political-economic developments, this kind of pressure has become virtually obsolete.

Chapter 2 points out the fact that skilled workers and subcontractors know what to do and how to do their work, but may need direction as to where, when and how fast to do it. Unskilled workers tend to move along with the job momentum.

Team workers tend to produce at a status quo level. Low performers are pulled up by the team, while high performers are held back. This topic is discussed further in Chapter 15, "Productivity and the Manning Chart".

The productivity of workers is not the superintendent's only concern; equal opportunity requirements also play a role in the contractual obligations between contractors and sponsors, especially when the sponsors are government agencies. The estimating and bidding phases should take into account any related costs, so that the superintendent can concentrate on the meeting of minority labor quotas, training and maintaining the labor force.

Legal, Ethical and Business Standards

The project contract documents contain specific directions concerning labor relations. In addition, the superintendent is committed to a variety of rules, regulations, and ethical and business standards in hiring, supervising, paying and discharging employees. Some of these considerations are listed below:

1. The Davis-Bacon Act states the contractor's obligation to pay prevailing wages and fringe benefits. Penalties are named for violations.
2. The Work Hours and Safety Standards Act defines overtime and penalties for non-payment.
3. A specified ratio is required of apprentices and trainees to journeymen where union trades are used.
4. Payroll records are required listing employees' names, addresses, work classifications, pay rates, fringe benefits, hours worked, and deductions made from paychecks.
5. Company contractual agreements with trade unions. They include:
    a. Pay rates and fringe benefits
    b. Hours of work
    c. Rest breaks
    d. Jurisdictions

Reference data, bulletins and manuals on current labor relations rules and regulations should be kept in the files in case of questions and disputes. Also, the names, addresses, and telephone numbers of labor authorities should be kept available so that clarifications can be obtained quickly.

Most contract documents require the posting of certain information and statements on the job site where they can be viewed easily by employees. Typical subject matter includes:

- Statement of non-segregated facilities (equal opportunity employment).
- Wage determination decision (Davis Bacon Act).

Most of the workers are temporary: that is, they are hired and then discharged as soon as they are no longer needed. Thus, the standard of workmanship is "average", and there is little incentive for workers to improve their performances. Faced with cost budgets, the superintendent is usually willing to employ the typical workers for the usual minimum periods of time. It is a temptation to keep superior workers or crews on the payroll, but a point is reached in the progress of the work when they cannot economically be maintained. The superintendent might then ask company management to assign these workers to another project or carry them on general overhead for a short time until real work materializes for them. This can be good long range strategy.

The superintendent can build up a fairly extensive file of workers of various skills. Since their abilities are known, they may be called upon at different stages of a project. In this manner, a team might also be carried over from one project to another.

Chapter 5
Sponsor Relations

The word "sponsor" is a blanket term. It covers any or all persons representing the person or entity paying for and eventually owning the real estate being developed by the contractor. The sponsor may be federal, state, county, or city government, or a private individual, and may be represented by civil service employees, military officers, engineers, architects, contract administrators, inspectors, or authorized persons carrying a variety of other titles.

The relationship between the superintendent and the sponsor's representatives is typically contractual. Nevertheless, unavoidable personality incompatibilities sometimes occur, and these can be damaging to both sponsor and contractor. If the superintendent has had a disastrous relationship with the same representative in a previous project, he might do well to make a written request for a change at the beginning of the project.

Normal differences of opinion are to be expected. There may be disagreements as to how strictly one should adhere to the letter of the contract, unless standards are set by others higher up in the sponsor's and contractor's hierarchies. Where standards have been predetermined, the superintendent and sponsor's representative merely carry out their employer's policies, and conflicts which cannot be resolved at field level are passed on up to the policy makers.

A second approach is one in which both sponsor and contractor let all project decisions and problems be resolved by their field representatives. These two approaches are the extreme. A third, more typical procedure is one in which the many minor decisions are made at field level, while the relatively few major decisions are made at a higher, administrative level. The dividing point between major and minor is usually decided by the superintendent and the sponsor's field representative on the basis of the cost, risk, aesthetic standard, or simply the inability to agree.

At the very beginning of the project, the superintendent might be able to judge the kind of relationship that will prevail. The policy and attitude of the sponsor might be known by reputation or from previous project experience. The sponsor's representative might frankly state his intentions, or the superintendent might receive signals indicating either a "hard-nosed" or a congenial approach.

A hard-nosed approach means that the sponsor will hold the superintendent to the strictest possible requirements and interpretations of the drawings and specs. He will not easily accept minimum standards of quality in materials and workmanship (as described in Chapter 19). The congenial approach means that the sponsor will be "reasonable" in this and other areas.

Sponsor relations are not only a factor in the contractor's bidding of a project, but also may affect the amount of the markup. As a general rule, an extra cost allowance is made in estimating and bidding if the sponsor has a hard-nosed reputation. It is management's responsibility to establish and carry out sponsor relations policies, but the superintendent must be aware of and work within these policies.

A hard-nosed approach taken by the sponsor is the short view. It disregards goodwill and repeat-business while concentrating only on achieving the maximum value from the present project. In the case of government projects, contracts must go to the lowest qualified bidder, there is little incentive for the sponsor to seek goodwill and repeat-business. Consequently, the hard-nosed approach is typical in these circumstances. The contractor for government projects also lacks the incentive to seek goodwill, since he cannot be denied future government contracts when he is the lowest qualified bidder. Drawings and specs for government projects tend to be exceptionally thorough, clear and standardized in order to minimize disputes. Even at their best, however, drawings and specs cannot achieve hair splitting accuracy. A gray area may exist between the interpretations of the sponsor and those of the superintendent. Conflicting interest and simple differences of opinion and values can widen the gap. A system of tolerances has evolved to address these issues.

For disputes that cannot be resolved by the parties involved, a complex system exists which tries to avoid work stoppages on the one hand and ensure payments on the other, by way of such sophisticated procedures and remedies as written directives, formal protests, arbitration, claims and lawsuits. Of these actions, the most immediate and within the superintendent's personal scope is the formal protest. When the sponsor's representative demands a material or performance which the superintendent believes is not in contract (NIC), or greatly exceeds minimum requirements, or places the contractor under risk of liability, the protest is an alternative to the unpopular and costly stop-work action. Figure 5.1 is an example of a letter of protest which would be sent to the owner's sponsor.

In a congenial relationship, letters of protest are used less often than in a hard-nosed relationship. For instance, a conflict such as the one illustrated in Figure 5.1 might be resolved at field level by a negotiated agreement for the sponsor to pay the extra material cost and the superintendent to do the installation work at no extra charge.

As with the hard-nosed approach, the congenial approach is also taken into account in the estimating and bidding stage. When the sponsor is known for congeniality, the bidder tends to lower prices and markup. This is a calculated risk, making it incumbent on the superintendent to further promote and maintain the congenial relationship. A good rapport has long range benefits for both sponsor and contractor and can lead to further business, including negotiated contracts.

The superintendent can improve and maintain any and all relationships by doing a superb job of the following:
- Keeping the quality of materials at or above the minimum accepted standards.
- Filing daily reports, especially those that may be required by the sponsor.
- Maintaining the construction progress schedule
- Carrying out quality control tests and inspections
- Submitting samples and shop drawings.
- Obtaining test certifications, etc.
- Policing the job in order to keep it clean and safe.

Date

Bell & Bell, Architectural Assoc.
100 Union Street
San Diego, California 92106

ATT: Mr. Chris Bell, Manager

RE: Administration Building & Fire Station
     Picacho, California
     Caulked Joints

Dear Mr. Bell:

Enclosed is a self-explanatory letter from the manufacturer of the specified material showing test results which indicate that caulked joints of the widths and depths detailed on the drawings are liable to fail from contraction forces when temperatures drop below 30 degrees. The manufacturer refuses to supply warranty unless the joint depths are increased 50%, at an additional cost of $1,800 (including installation labor).

This proposal has been rejected by your field engineer; therefore, we will proceed in accordance with the drawing details under protest and without warranty of responsibility for joint failure.

                              Very truly yours,

                              Your Own Construction Company

                              B. Buttress
                              Superintendent

Encl.

Figure 5.1

It was mentioned at the beginning of this section that the sponsor-superintendent relationship is basically contractual and financial. The power to control the purse strings is a major management tool possessed by the sponsor. Techniques for limiting the potential abuses in both billings and payments are discussed in the next section.

Selling the Job

Daily communication between the superintendent and the sponsor's representative is guided by schedules, procedures, and reports (as described in Chapter 10). When the project is completed on the scheduled date, only the selling of the job (persuading the sponsor to accept it as totally completed) remains. This should be only a matter of correcting a number of small imperfections.

There should be only one walkthrough (final inspection) and only one "punch list". The superintendent can enlist the aid of the sponsor's field representative as well as his own company's manager in requesting the attendance of every authorized person at the walkthrough. It is also a good idea to invite representatives of the major subcontractors. If one or more representatives cannot participate on the appointed day, it might be best to reschedule the walkthough in order to avoid second or third tours with resulting additions to the punch list.

The major subcontractors can show and describe operational equipment, such as pumps, air conditioners, heaters, and so forth. They can also answer and explain technical questions directed to their areas of specialization. At this point, they can make time commitments for expediting punch list items.

Because it is the final inspection, the walkthrough is the most critical moment faced by the superintendent. To minimize the criticisms and shorten the punch list, the superintendent and sponsor's representative can make their own preliminary walkthrough, and the superintendent can expedite the correction of many anticipated problems.

As a rule, the sponsor is eager to obtain possession and use of the long-awaited facility and is willing to accept the project upon completion of the punch list work. Goodwill between contractor and sponsor, maintained throughout the project, is also influential in selling the job. In consideration of contractor warranties, the sponsor is inclined to agree on important issues and to waive or postpone the correction of trivial points. Figure 5.2 is an example of a punch list with a report of actions to correct criticized items.

When all items on the punch list are corrected to the satisfaction of the sponsor, the way is cleared for three last events: (1) contractor move-off and final cleaning of the site, (2) final payment of the contractor, and (3) sponsor move-in. It is at this time that the sponsor is given the key to the facility.

During the warranty period (typically one year), the superintendent (with his intimate knowledge of the project), might occasionally be called on to analyze problems and arrange for their correction.

PUNCH LIST

PROJECT: Admin. Bldg. & F.S.   **DATE OF WALKTHROUGH** 10/1/87

TIME START 9:15 AM   **TIME COMPLETE** 12:00 Noon

REPRESENTING SPONSOR:

Chris Bell, Architect

Chuck Bloomfield, Engineer

N. Neilson, Inspector

REPRESENTING CONTRACTOR:

Buck Buttress, Superintendent

U.N. Level, Foreman

S.O. Beit, Manager

Doug Piper, Piper Piping Co.

E.C. Hull, Sparks Electric

ITEMS REQUIRING CORRECTION

1. M.H. cover too high at station 14+10

2. Entrance door closer needs adjusting

3. Sprinkler piping not painted

4. Sheet metal flashing not straight

5. Landscaping not completed

6. Parking light doesn't turn on

7. Rollup door needs adjusting

8. Robe hook missing in men's room

CORRECTIVE ACTION

1. Piper Piping Co. will lower to proper elevation

2. Foreman will adjust

3. Painter was notified

4. Sub was notified

5. Sub promises completion by Friday

6. Elec. sub is checking it

7. Sub was notified

8. Foreman will supply

Figure 5.2

Part II

Managing the Project

Chapter 6
Planning the Working Area

Space for offices, trailers, sheds, materials, and equipment is usually restricted on the job site, and careful planning is necessary. Even when space is not a problem, wise planning brings the rewards of efficiency and economy.

Example 1: A multi-story building (Figure 6.1) will cover the entire site; there is no free adjoining property available for working area. The following possibilities exist:

1. Annex, with city approval, portions of adjacent streets for pedestrian walkways and working areas.
2. Construct foundations and a floor slab for the ground floor auto parking garage; use this space for the working area while erecting the building overhead.
3. Make all temporary facilities as portable as possible, and move them about on the site as needed to clear areas for construction work.
4. Rent space as near to the project as possible.
5. Use a combination of the above four methods.

Example 2: A large site (Figure 6.2) with two adjoining streets, provides the opportunity for efficient traffic control and for the flow of construction equipment (such as cranes and concrete trucks) around all sides of the new building. Because of the position of the office and storage shed, a minimum of security fencing is required around the construction yard.

Example 3: A project scattered over several isolated sites may be more efficiently served by two or more construction yards with primary and secondary field offices, as in Figure 6.3. Communication and transportation are important between sites; thus, two or more telephones and walkie-talkies may be a worthwhile investment. A planned travel route between sites (for all personnel and equipment) might economize on mileage and time.

The superintendent is committed to the budgets (proposed in the bidding stage) for such items as field office, personnel, telephone, fences, storage sheds, and toilet facilities. These items are usually listed in the contract documents under the heading "General Conditions" (see Chapter 13—"Cost Controls"). Figure 6.4 shows a filled-in form listing sample estimates of general conditions prices. In this section, discussion is confined to those items identified by an asterisk in the left-hand column. Note that the cost budgets cover quantities and unit costs only; they do not specify details such as type of construction and materials. The superintendent has some freedom of choice where temporary facilities are concerned.

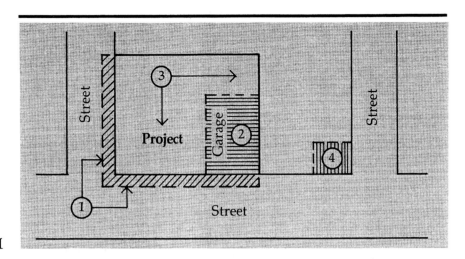

Figure 6.1

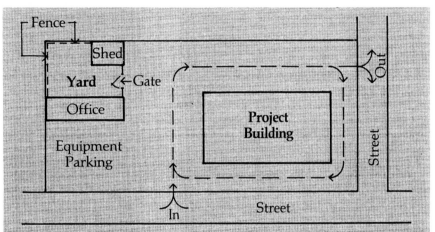

Figure 6.2

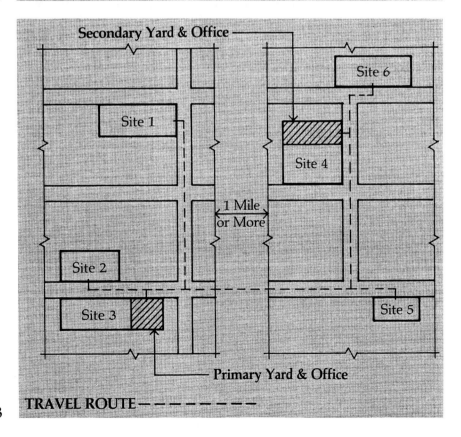

Figure 6.3

Means Forms
COST ANALYSIS

PROJECT: Admistration Bldg. & Fire Station	SHEET NO
LOCATION: Acacho, CA	ESTIMATE NO
TAKE OFF BY: PC QUANTITIES BY: PRICES BY: PC	CLASSIFICATION ARCHITECT
PRICES BY: PC EXTENSIONS BY AT PRICES BY AT	DATE 10/9/87 CHECKED BY: JM

Item	Quantity	Unit	Material Unit Cost	Material Total	Labor Unit Cost	Labor Total	Equipment Unit Cost	Equipment Total	Subcontract Unit Cost	Subcontract Total	Total
General Conditions											
1. Layout Structures	4	Days			625.	2500					2500
* 2. Field office	12	Mo.	118.	1416							1416
* 3. Storage Trailer	12	Mo.	77.	924							924
* 4. Toilets	24	Mo.	73.	1752							1752
* 5. Telephone	12	Mo.	150.	1800							1800
* 6.a Water Hook-Up-Meter	—	L.S.		300		300				1500	2100
b " monthly Charge	12	Mo.	60.	720							720
* 7.a Electrical Service	—	L.S.		400		500				1268	2168
b " monthly Charge	12	Mo.	70.	840							840
* 8. Security Fence	400	L.F.							9.05	3620	3620
* 9. Signs	—	L.S.								250	250
10. Temp. Weather Closures	—	L.S.		100		800					1050
11. Office Supplies	12	Mo.	35.	420							420
12. Small Tools	12	Mo.	200.	2400							2400
13. Misc. Rental Equip.	12	Mo.	400.	4800							4800
14. Oil, Fuel, Maintenance	12	Mo.	250.	3000							3000
15.a Clean Up, Progressive	12	Mo.	100.	1200		14400				720	16320
b " , Final	240	Hrs.	5.	1200							5064
16.a Superintendent pay	12	Mo.				44880					44880
b Foreman	6	Mo.				20208					20208
c Clerk	12	Mo.				9600					9600
17.a Barricade at Street	700	L.F.	6.	4200		4200					8400
b " Night Lights	12	Mo.	60.	720							720
18.a Miscellaneous	—	L.S.		1000							1420
Sub-Totals				27442		101772				7358	136572

Figure 6.4

22

This phase is often called "the project set up", or "mobilization". "Plant" is another term for the temporary facility and includes the cost of supervisory and clerical personnel required to administrate and coordinate the construction work. The plant is moved onto the site, assembled, erected, maintained, disassembled, and moved off the site (mobilized and demobilized).

The field office could be any weatherproof structure of adequate size. Because of portability and reusability, a house (office) trailer is commonly chosen by builders. Figure 6.5 is the floor plan of a typical office trailer, providing space for the superintendent and secretary or time clerk in section A, and the sponsor's inspector or a quality control representative in section B. In the event that section B is not used as an office, it may be used for storage of small tools or hardware; or it might serve as a conference room. The plan table area should be available to foremen and subcontractors. An interior bulletin/tack board serves the superintendent's needs, and another board outside directs notices to the workers (see Chapter 4—"Labor Relations"). Drinking water is available both inside and outside the office, and a hose bibb provides water for miscellaneous purposes. Interior rest rooms are available to the office staff and visitors whenever sewer connections exist on the site. Other features include a burglar alarm system, night lights for security, and a locked, fireproof safe.

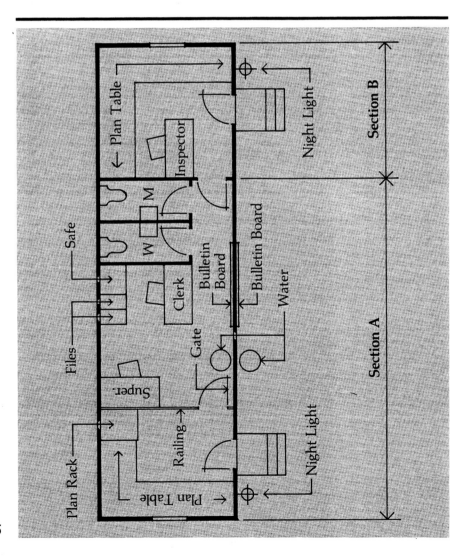

Figure 6.5

If the office is owned by the contractor, its cost may not concern the superintendent, since the budget of $118.00 per month is then only an allowance for depreciation. However, repair and maintenance of the office is still a consideration. The office may be classified as a piece of "equipment" under item 14 (see Chapter 20—"General Conditions").

The storage shed in Figure 6.4, item 3, represents a cost allowance, presumably for one small structure. The superintendent must decide the number, sizes, and types of structures required for the project. A mobile shed, rented or owned by the company, will store small tools, hardware, paint, and accessories. A roofed structure may have to be constructed to protect such items as lumber, plywood, doors, milled trim and cement. Roofed bins may also be needed to store gravel, sand, and other bulk materials and to protect them from the rain. It is often more economical to use tarpaulins or plastic coverings, particularly when construction yard space is limited.

Custom-built sheds may be prefabricated for reuse in future projects. In this case, the materials are prorated rather than charged entirely to one project. The superintendent should clarify such expenses with the bookkeeper to stay within the budget. An example is a new 20' x 40' shed designed to be easily assembled and disassembled for reuse in future projects. The cost is:

Material	$2,000	5 uses =	$ 400
Labor-prefab	$1,200	5 uses =	$ 240
assemble	$ 700		$ 700
disassemble	$ 600		$ 600
	$4,500		$1,940

Temporary Toilets are usually portable chemical units. The budget has provided for two of them at a rental rate of $73 per month each (including move on and off and maintenance).

Temporary Telephone Service, installation and monthly charges are lumped together in item #5 of the budget at an average cost of $150 per month. When installation and service charges are prorated and deducted, the superintendent may find that he has, say, $100 per month remaining to cover long distance calls. Communication is vital to the progress and economy of a project and, therefore, use of the telephone should not be miserly. Nevertheless, a conscious effort is needed to avoid unauthorized usage by subcontractors, workmen, and others. To this end, the superintendent might instruct the clerk to keep a log of all toll calls and to backcharge users (see Chapter 18, "Progress Payment Requests and Backcharging"). A simpler alternative might be the installation of a pay phone for the subcontractors' use. This phone is usually mounted outside the superintendent's trailer so that it is accessible even if the superintendent is not there.

Temporary Water: The water needs of a project fall into three classes: (1) drinking water, (2) general purpose water and (3) bulk water for excavation or other purposes. The budget item #6a (Figure 6.4) provides a total of $2,100 for all piping, meter, and connection to the nearest public source. Only general purpose water and drinking water are provided for. Bulk water is provided by the excavation subcontractor. The superintendent negotiates (Figure 6.6) with the mechanical subcontractor for the installation of temporary water service, meter, hose bibbs and other accessories (see Chapter 17—"Letting Subcontracts"), for the lump sum of $1500, leaving $600 for bottled drinking water service.

```
                        WORK ORDER
                       (Subcontract)


PROJECT  Admin. Bldg. & Fire St, Picacho, CA  DATE    11/12/87


       Piper Piping Co.                          agrees to furnish all
labor, material and equipment necessary to:

    Install temporary water piping from nearest

    domestic water connection underground to the

    office; a riser in the construction yard; and

    a riser where directed at the construction

    site.




                                                    in accordance with
the drawings and specifications for the sum of:

One thousand five hundred dollars  ($1,500.00              ).

To be completed by: November 11, 1987

INCLUDED: Water meter, PVC piping, hose bibbs,

   misc. fittings and connection to public main.



FOR:                            YOUR OWN COMPANY

BY: Doug Piper, Mgr.            BY: Buck Buttress, Super
```

Figure 6.6

Temporary Electrical Service is negotiated with the electrical subcontractor for a sum of $1,800. (Figure 6.7 is a typical work order written in the field for work not included in the basic subcontract.) It is helpful if the superintendent has company authorization for such incidental work orders. Chapter 17, "Letting Subcontracts and Processing Change Orders", addresses this aspect of contract writing. A pole is provided in the construction yard for power to the office and to one or two points on the site convenient to a variety of electrical tools, such as concrete vibrators, finishers, circular saws, impact hammers, and drills. $368.00 remain in the budget for miscellaneous accessories, such as cords, outlets, switches, and bulbs.

A Temporary Security Fence around the construction yard also requires a work order written for a specialist. 400 linear feet of fence at $9.05 per foot was included in the budget (item # 8). The superintendent may find that in actuality, only 320 linear feet are required, and might negotiate a unit price of $8.80/lf, as shown in the work order (Figure 6.8).

Temporary Signs will consist of one large project identification board with professionally printed names of the sponsor, the architect and engineer, the prime contractor, and the major subcontractors. In addition, the superintendent decides to place an "entrance" and "exit" sign at chosen access points. He will assign the construction of the frames and posts to a carpenter and sublet the printing as specified in Figure 6.9.

A pedestrian barricade is required, and budget item 17a (Figure 6.4) provides $8,400. Item 17b provides $720 for lighting the barricade at night. The superintendent designs a suitable barricade of lumber and plywood construction and makes the following estimate of its cost:

Lumber (used)	8,000 bf @ .17 =	$1,360
Plywood (used)	8,400 sf @ .15 =	$1,260
Labor-construct	16,400 bf @ .20 =	$3,280
Labor-dismantle	16,400 bf @ .10 =	$1,640
Cleanup & haul to storage	16,400 bf @ .07 =	$1,148
		$8,688
Less salvage value	$16,400 bf @ .05 =	$ 820
	Total	**$7,868**

The superintendent plans to string incandescent lights along the barricade, attaching them to one of the risers provided by the electrical subcontractor.

Expected cost of lighting is:

Material	$275
Monthly charges 12 @ $16.00	$192
	$467

Total savings are expected to be:
$(8,400 + 720) - (7,868 + 467) =$ $785

For tips on cost estimating, see Chapters 17 and 18.

```
                    WORK ORDER
                   (Subcontract)


PROJECT Admin. Bldg, & Fire Sta., Picacho, DATE  11/12/87
                          CA
  Sparks Electric Co.              agrees to furnish all
labor, material and equipment necessary to:
   Provide a pole at the construction yard,
   110/220 v wiring into the office, meter, and
   two risers with receptacles where directed
   on the site.
_____
_____
_____
_____
_____
_____ in accordance with
the drawings and specifications for the sum of:
One thousand eight hundred dollars ( $ 1,800.00        ).
To be completed by: November 21, 1987
INCLUDED: Transformer
_____
_____
_____

FOR:                         YOUR OWN COMPANY
BY: E.C. Hull, V.P.          BY: Buck Buttress, Super
```

Figure 6.7

```
                        WORK ORDER
                       (Subcontract)


PROJECT Admin. Bldg. & Fire Sta., Picacho, CA DATE  11/13/87

 Rental Fence Co.                            agrees to furnish all
labor, material and equipment necessary to:

   Install 320 LF of 6 foot high chain link
   fence with 2 strands of barbed wire
   around the construction yard.


_____

_____

_____

_____

_____

_____
                                    in accordance with
the drawings and specifications for the sum of:

 Two thousand eight hundred sixteen     ( $ 2,816.00        ).
                              dollars
To be completed by: November 20, 1987

INCLUDED:  1 pair of gates 12' wide; removal of
 fencing upon completion of project

_____

_____


FOR:                           YOUR OWN COMPANY

BY: Geo. Picket, Sales         BY: Buck Buttress, Super
```

Figure 6.8

WORK ORDER
(Subcontract)

PROJECT <u>Admin. Bldg. & Fire Sta, Picacho, CA</u> DATE <u>11/13/87</u>

<u>Neat Signs, Inc.</u> agrees to furnish all
labor, material and equipment necessary to:

<u>Paint the lettering on sign boards, furnished</u>
<u>by the contractor, in accordance with</u>
<u>the sketches attached hereto.</u>

_____

_____

_____

_____

_____

_____ in accordance with
the drawings and specifications for the sum of:

<u>Two hundred twenty-eight dollars</u> (<u>$228.00</u>).

To be completed by: <u>November 20, 1987</u>

EXCLUDED: <u>Sign boards and frames; pickup</u>
<u>and delivery</u>

_____

_____

FOR: YOUR OWN COMPANY

BY: <u>Ned Neat, Pres.</u>   BY: <u>Buck Buttress, Super.</u>

Figure 6.9

Chapter 7

Project Security

A construction project is particularly vulnerable to theft and vandalism because it stands vacant during precisely those hours when criminals are most active. Security provisions can be costly, and this responsibility generally falls to the superintendent. Following is a list of basic security measures that can be taken to avoid loss or damage. They are listed roughly in order from the most effective to the least.

1. One or more full-time guards.
2. Electrically charged security fence and locked gate.
3. Security fence (not electric), locked gate, and guard dog.
4. Patrol guard service and burglar alarm system
5. Security fence and burglar alarm system.
6. Burglar alarm system only.
7. Fence only.
8. Lights only.

Without adequate budget provisions, the first four methods may have to be ruled out; but one or a combination of the remaining measures might cost less in the long run than losses due to leaving the project completely unprotected.

The type of construction is a guide to the appropriate method of security. For instance, fireproof construction is less vulnerable than wood frame construction. Concrete construction is less vulnerable than light steel framing.

The state of completion also influences the method of security. In the first few weeks or months, there are few materials on the site; in the last few weeks or months, a building might be sufficiently completed to be locked up and/or its openings barricaded. The middle months involve the greatest vulnerability, thus the cost of security may be concentrated in that period.

Security is a consideration not only when the project stands vacant, but also during the course of daily construction activities. Secure areas should therefore be provided for items that, while small and not necessarily expensive on an individual basis, are very important to the overall project schedule. Door frames, rough hardware, special blocking or inserts are examples. If such components are misplaced or stolen, the schedule could be adversely affected. Thus, a small cost item, improperly handled, could have a large cost effect.

Throughout the course of the project, subcontractors usually maintain tight security regarding their own important items. The protection they need is generally provided by on-site, locked storage trailers. The foreman of each subcontractor is responsible for trailer access. As the project progresses and it becomes necessary to remove the trailers, storage rooms within the structure may be utilized. Larger tools (such as compressors, generators, pumps, highlifts or tool gang boxes) that cannot easily be moved by one man may be left out on the project site, but are usually secured with a chain and lock to an immovable object. Extremely heavy equipment such as site work equipment, cranes, etc. are generally parked in a conspicuously lighted, easily viewed area in order to deter malicious vandalism.

Chapter 8

Environmental Considerations

Many specifications provide detailed instructions as to the use and care of the construction site, as it affects the surrounding land, air, water, plants, animals and human beings. If cost is involved in environmental measures, appropriate budgets are provided by the estimator prior to bidding. Specified environmental protection instructions are mandatory and must be carried out by the superintendent as conscientiously as any other portion of the project. In addition, certain protective measures should be regarded as standard practice, even if they are not mentioned in the specs. Some typical environmental protection measures involve the following procedures:

- Avoiding, fencing, or isolating endangered species of plants and animals.
- Constructing grades to control water run-off.
- Controlling erosion with ground cover or other means.
- Preventing dust by wetting down, barricading, etc.
- Properly disposing of various substances, such as toxic chemicals.
- Suppressing noise.
- Suppressing odor.

All environmental control measures should be logged, either on the job log (see Chapter 10—"Record Keeping") or in a special environmental records file. In addition to written documentation, photographs should be taken whenever possible. With changing seasons and the progress of the project, an area of the project can look completely different to the untrained eye. The rules and regulations of environmental impact items can be so intricate and complex that misunderstandings often develop even among parties well versed in this subject. It is extremely worthwhile to develop a congenial relationship with members of municipal commissions involved in the project. A good rapport can be encouraged by attending local meetings, getting to know the people on the commission, and keeping the commission well informed—especially on the more delicate or controversial issues.

Chapter 9

Introduction to the Project

Following successful bidding, the bid package is turned over to the superintendent in a process called "briefing the super". It is at this time that the field project manager, or superintendent, takes charge of the project. Along with the bid package, he also acquires some essential information and reference data, such as:

- A complete set of drawings and specs with addenda marked "Official Field Set" (see Chapter 6, "Planning the Working Area").
- Copies of all subcontracts and purchase orders to date.
- Copies of estimates, budgets, and cost control forms.
- Oral descriptions (and illustrations) of the construction methods upon which the bid was based.
- Progress schedules showing the time periods allotted to each trade and work item.
- Format and data for requesting monthly progress payments.
- General information as included in the Superintendent's Information/ Instructions sheet (see Figure 9.3).

The drawings and specs, together with a site inspection, provide most of the visual information. The number of structures to be built, their shapes, dimensions, heights, and locations on the site are envisioned. Areas to be paved and/or landscaped are identified. Sources of water and power needed for the construction work are noted. The extent of earthmoving and surface water run-off are roughly conceived. The superintendent studies the options for construction yard and office locations (see Chapter 6, "Planning the Working Area").

All information pertaining to the subsurface should be studied for the identification and protection of buried utility lines (water, gas, sewer, electrical, steam, etc.), and for the possibility of water requiring pumping or drainage. Unusual rock, soil, or mineral conditions should also be considered. Much of this information will have been researched previously by the architect, engineer, and others and might be described in the drawings and specs. The superintendent will want to confirm the given information, pinpoint locations and discover any additional facts about the site that might have been overlooked.

Subcontracts and purchase orders should accurately define the scope of each subcontractor's and supplier's contribution to the project. When studying these agreements, the superintendent seeks the answers to such questions as:

- Specifically what trades and items of work are included, and what work is excluded?

- Are items of work excluded from one subcontract included in some other subcontract?
- If there are any items not covered by subcontract or purchase order, who is properly responsible for them?
- Is any item included in more than one subcontract? If so, one of the subs should be asked to delete it, at a negotiated deductive price.
- Do the subcontracts contain specific agreements regarding responsibility for such general conditions items as scaffolding, barricades, safety provisions, and clean up?

Any omissions, double-coverages, conflicts, or ambiguities discovered in the agreements should be investigated and resolved early in order to prevent costly delays. The drawings, specs, and subcontracts should be screened for questions that warrant investigation. Some of these questions might be answered directly by the expeditor (formerly the person in charge of the bidding). As the superintendent seeks answers to the remaining questions, he has the opportunity to meet with the subcontractor's representatives. Introductory conversations with every subcontractor and supplier might raise still more questions, but if these issues are resolved early, progress should be smoother during construction.

Another method for visualizing the project and seeing that all elements are accounted for is to use a horizontal spread sheet similar to a CPM/PERT system. Figure 9.1 is an example of such an analysis sheet. Pinned to the field office tack board, it provides quick reference for all interested persons. In this example, certain items are numbered to focus the superintendent's attention on the fact that they require his arrangements. All other work is the responsibility of subcontractors, but the superintendent needs to identify the point of separation between the trades. For instance, 10 gauge thickness of metal is a division point between the sheet metal and the miscellaneous metal subcontractors (it is also possible that both subcontractors might exclude 10 gauge metal, or both include it).

The trades listed horizontally in the center of the sheet represent individual subcontracts. The material and work items that appear above the center clarify subcontract responsibility. Items below the center are not included in subcontracts; provisions for their disposition are noted.

Only those materials or work items that merit special note are listed. It is understood that all other items are clearly part of subcontractor trades and general contractor work either by spec section or custom-of-the-trades.

Estimates, budgets, and cost control forms will be fully discussed in Chapter 13, "Cost Controls".

During the bidding phase, every item in the project was assigned a dollar value; to the superintendent these values are budgets to which his job commits him. The effort to keep within each and every assigned dollar value tends to increase the overall profit, or minimize the loss from items that might have been estimated too low.

Good cost records help the estimating department produce realistic budgets for future projects. For this reason, it is important that the superintendent keep an accurate and conscientious accounting. Understandably, it is a matter of pride to the superintendent to show cost records consistent with, or under the budget. This feat is sometimes accomplished by borrowing and lending between underrun and overrun cost items. An example is shown in Figure 9.2 where the correct record shows all unit costs except formwork exceeding the estimated budgets. It is possible to distort the records so that all unit costs appear lower than their respective budgets, but this kind of record keeping is

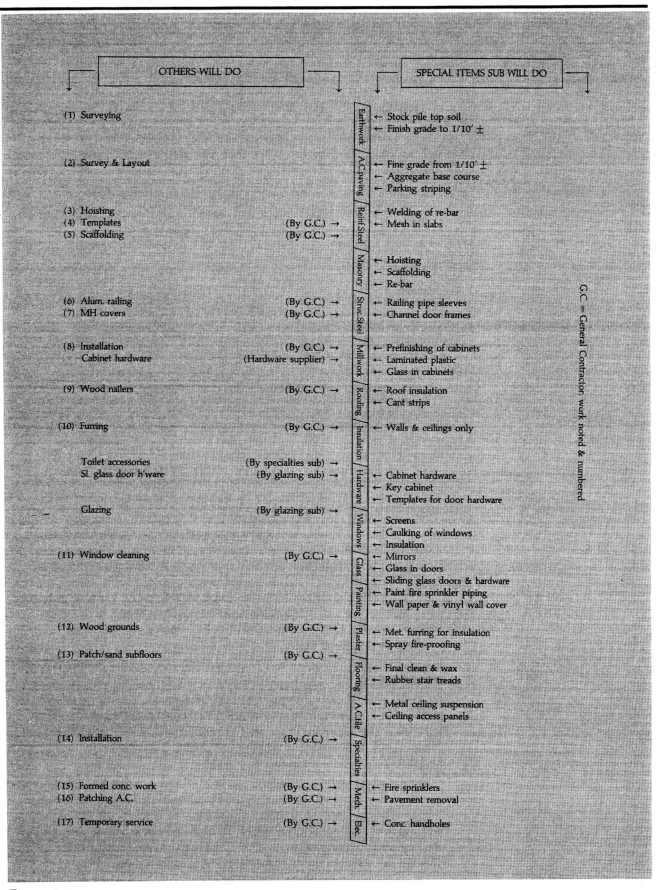

OTHERS WILL DO			SPECIAL ITEMS SUB WILL DO

(1) Surveying		Earthwork	← Stock pile top soil
			← Finish grade to 1/10′ ±
(2) Survey & Layout		A.C.paving	← Fine grade from 1/10′ ±
			← Aggregate base course
			← Parking striping
(3) Hoisting		Reinf.Steel	← Welding of re-bar
(4) Templates	(By G.C.) →		← Mesh in slabs
(5) Scaffolding	(By G.C.) →		
		Masonry	← Hoisting
			← Scaffolding
			← Re-bar
(6) Alum. railing	(By G.C.) →	Struc.Steel	← Railing pipe sleeves
(7) MH covers	(By G.C.) →		← Channel door frames
(8) Installation	(By G.C.) →	Millwork	← Prefinishing of cabinets
Cabinet hardware	(Hardware supplier) →		← Laminated plastic
			← Glass in cabinets
(9) Wood nailers	(By G.C.) →	Roofing	← Roof insulation
			← Cant strips
(10) Furring	(By G.C.) →	Insulation	← Walls & ceilings only
Toilet accessories	(By specialties sub) →	Hardware	← Cabinet hardware
Sl. glass door h'ware	(By glazing sub) →		← Key cabinet
			← Templates for door hardware
Glazing	(By glazing sub) →	Windows	← Screens
			← Caulking of windows
			← Insulation
(11) Window cleaning	(By G.C.) →	Glass	← Mirrors
			← Glass in doors
			← Sliding glass doors & hardware
		Painting	← Paint fire sprinkler piping
			← Wall paper & vinyl wall cover
(12) Wood grounds	(By G.C.) →	Plaster	← Met. furring for insulation
			← Spray fire-proofing
(13) Patch/sand subfloors	(By G.C.) →	Flooring	← Final clean & wax
			← Rubber stair treads
		A.C.tile	← Metal ceiling suspension
			← Ceiling access panels
(14) Installation	(By G.C.) →	Specialties	
(15) Formed conc. work	(By G.C.) →	Mech.	← Fire sprinklers
(16) Patching A.C.	(By G.C.) →		← Pavement removal
(17) Temporary service	(By G.C.) →	Elec.	← Conc. handholes

G.C. = General Contractor; work noted & numbered

Figure 9.1

Partial Cost Record
(Labor and equipment only)

A. Correct Record

		Estimated			Actual to Date	
		Quantity	U.C.	Quantity	Total Cost	U.C.
Footing	Exc.-mach.	440 cy	4.07	455 cy	2,366	5.20
	Exc.-hand	60 cy	16.10	53 cy	1,120	21.13
	Backfill	200 cy	7.15	198 cy	1,723	8.70
	Formwork	9,200 sfca	3.06	9,200 sfca	22,632	2.46
	Concrete	300 cy	50.90	305 cy	18,575	60.90
					46,416	

B. Distorted Record

		Estimated			Actual to Date	
		Quantity	U.C.	Quantity	Total Cost	U.C.
Footing	Exc.-mach.	440 cy	4.06	455 cy	2,193	4.82
	Exc.-hand	60 cy	16.10	53 cy	928	17.51
	Backfill	200 cy	7.15	198 cy	1,376	6.95
	Formwork	9,200 sfca	3.06	9,200 sfca	23,869	2.59
	Concrete	300 cy	50.90	305 cy	18,050	59.18
					46,416	

Figure 9.2

misleading and might cause the company estimator to under- or over-bid a future project. The superintendent does well to maintain an impersonal, objective attitude toward cost records.

When the superintendent is first introduced to a project, cost record forms and the data they contain help to inform him. The construction methods conceived by the estimator and used as a basis for bidding the project are similar to cost budgets in that they set a maximum standard for the superintendent. The cost is only one consideration; others are time (which is also a cost factor) and safety. The superintendent's attempt to recognize potential construction problems and available methods is part of his introduction to the project.

It is possible that one or more of the methods suggested in the bid package will be challenged by the superintendent as unsafe, and alternate procedures recommended. It is his prerogative to devise superior methods. Generally, the term "safety" includes a regard not only for human beings, but also for materials and equipment. There are few considerations in the construction business that do not have cost value, and safety is no exception. (See Chapter 19 for more on safety precautions.)

Progress schedules are sometimes given to the superintendent as part of the project package and are a good source of information. Ideally, the progress schedule is only tentative, and the superintendent is given the opportunity to draw up, or at least contribute to the final and official schedule. The detailed drawing up of a progress schedule is often assigned to specialists. See Chapter 14—"Time Controls" for further discussion of this subject.

The progress payment request master breakdown is often given to the superintendent as a part of the project package. Ideally, the superintendent participates in developing the system. In any case, the payment request forms are a source of in-depth information for the superintendent. See Chapter 18—"Progress Payment Requests" for further discussion of this subject.

The superintendent's information/instruction sheet is provided by some companies as a part of the bid package. It is directed to the main points of interest and provides information on materials, sources, subtrades, distribution of duties and other matters which were decided in the bidding phase of the project. The superintendent may also be given a copy of the purchase order or some other description of the work or products that each subcontractor or supplier is expected to do or supply. Figure 9.3 is an example form, filled in to show important data. Such a form typically includes:

- The name of the project.
- Proposed starting and completion dates.
- Names and telephone numbers of the architect, engineer and representatives.
- Names and telephone numbers of the sponsor and representatives.
- Names and numbers of subcontractors and their representatives.
- Names and telephone numbers of material suppliers and their representatives.
- Items and responsibilities delegated to the superintendent.
- Items and responsibilities delegated to the office staff and others.
- Special information and instructions that the superintendent should have.

This list is not intended to be final; the superintendent may suggest changes and improvements. Its purpose is to introduce him to valuable details accumulated by the bidding team and to enable him to carry out the construction work with as much autonomy as possible. The following list is an example of the kind of general information and guidelines that may be passed along to the superintendent prior to the start of the job.

The superintendent might be informed of his responsibility for the following activities:

- Consult with the engineer designing the progress schedule.
- Design and build the pedestrian barricade.
- Maximize local hiring of workmen to minimize subsistence costs.
- Arrange for temporary water and power services.
- Consult with engineer designing earth shoring.
- Search for more economical sources of import fill.
- Note the list of loose ends and help resolve them.
- Arrange for field office, storage shed, toilets, security fence, telephone, project sign and other temporary facilities.
- Search for sources of heavy earthmoving rental equipment, and supervise the grading work.
- Provide shoring for all openings in masonry walls.
- Provide hoisting for rebar to second floor and above.
- Provide formed concrete work for mechanical trades.
- Provide cant strips for roofing sub.
- Provide concrete base for flagpole.
- Directly purchase small quantities of materials as needed.

The Office will be responsible for the following:

- Furnish a progress schedule (see subcontract list).
- Furnish a cost record keeping program.
- Order large quantities of materials through purchase orders.
- Perform general expediting of subcontractors' submittals.

SUPERINTENDENT'S INFORMATION/INSTRUCTIONS SHEET
(Page 1 of 4)

FOR PROJECT: ___Administration Bldg. & Fire Station, Picacho___

GENERAL INFORMATION:

STARTING DATE: _Nov. 5, 1987_   COMPLETION DATE:

OFFICIAL: _Nov. 5, 1988_

PROPOSED: _Oct. 5, 1988_

A & E: ___Bell & Bell Architects___

REPRESENTATIVE: __Chris Bell__   PHONE: ___777-7777___

SPONSOR: ___City of Picacho, California___   ___737-7373___

REPRESENTATIVES:

CONTRACT ADMINISTRATOR_____

BUILDING INSPECTOR_____

PROJECT INSPECTOR_____not known_____

SUBCONTRACTORS:

TRADE	SUB	REP	PHONE
Millwork	Johnson's	Sawyer	737-7373
Calking	Liners	Bill	544-5444
Surveying	Sight Bros.	O.O. Sight	555-5552
CPM schedule	Charters	Diane	737-4872
Chain link fence	Smith Bros.	?	297-7414
A. C. Paving	Blazers	T. Blazer	554-2496
Landscaping	Greener	Joe	676-6766
Reinforcing steel	Betbar	?	474-8888
Masonry	Histack	Curtis	757-0930
Struc. steel/misc. met.	Ringer	Paul	545-2200
Steel decking	Dekko	Phil E.	676-1084
Wall & clg, insulation	Comfy	Jean	297-7077
Roof insulation	Ample	?	554-5554
Clay tile roofing	Tiletop	Swanson	676-1111
Sheet metal	Leaders	Al Barnes	345-5000
H.M. doors & frames	Doors Unltd.	Johnson	666-6666
Coiling steel doors	Steelco	Barbara	336-5678

Figure 9.3

The following special information and instructions might be included:

- All costs of testing and inspection will be borne by the sponsor, except for concrete compression tests and mix designs.
- If possible, excavate footings neat (no side forms).
- Eliminate disposal of excess footing excavation material by using as site fill. This will also save a small amount of import material.
- Arrange for installation of bath accessories by company employees. Subs will install all other building specialties.

Chapter 10
Record Keeping

The subject of paperwork can be controversial among superintendents. The optimum amount of record keeping is difficult to establish, but there is rarely an overabundance in construction management. On the contrary, the problem occurs when the records fall short of company needs. The challenge is finding the time for this important task. Partial relief has appeared in recent years with the use of computers (see "Computers in the Field", later in this chapter). There are also some basic forms that can be used to quickly and effectively record important information. Following are descriptions and explanations of these reports.

Daily Reports

These documents serve two main purposes: (1) they furnish information to off-site persons who need and have a right to know important details of events as they occur daily and hourly, and (2) they furnish historical documentation that might later have legal bearing in cases of disputes.

Drawing upon years of experience, contractors have developed certain Daily Report forms to guide superintendents in the kinds of information and data worthy of recording. Daily reports should be as factual and impersonal as possible, free from the expression of personal opinions and feelings. Figure 10.1 is an example of a daily report which is essentially a public document. Comments of a more personal nature should be recorded in diaries. Factual information not suitable for daily reports or diaries should be recorded in logs.

It is important that daily reports be numbered so that they are easily kept in order. The numbers should correspond faithfully with the working days established on the progress schedule, and with the manning chart. In this way, comparisons can be made easily. In the event of no-work days, a daily report should still be made, stating "no work today" (due to rain, strike, or other causes).

A description of the weather can be of value as it justifies lagging progress, the need for overtime work, time extensions, or nonbudgeted expenses, such as dewatering, waterproofing, wind bracing and repairing of damage. Temperature records can also explain variations in design mixtures like concrete. The temperature factor may help to account for the quantities placed, early or late hours of work, and cost overruns (extremes of temperature may affect production).

The direct work force is composed of the superintendent's company employees. A daily record of the total number of employees is useful for a variety of reasons, including its effect on insurance coverage. Keeping track of the number of workers of different classifications is also useful. By comparing and confirming these numbers to the manning chart (see Chapter 15— "Productivity and the Manning Chart"), an optimum number of workers can be determined.

Figure 10.1

Means Forms

DAILY CONSTRUCTION REPORT

JOB NO. 112

PROJECT: Admin. Bldg. & Fire Station
ARCHITECT: Picacho, CA

DATE: Thursday, 4/10/87
SUBMITTED BY: Buck
WEATHER: Showers & Windy   TEMPERATURE: AM 62 PM 75

CODE NO	WORK CLASSIFICATION	FOREMEN	MECHANICS	LABORERS	SUB CONTROL	TOTAL HOURS	DESCRIPTION OF WORK
	General Conditions	4				32	Super & Foremen, Clerk
	Site Work: Demolition						
	Excavation & Dewatering						
	Caissons & Piling						
	Drainage & Utilities						
	Roads, Walks & Landscaping						
	Surveying	3				24	
	Concrete: Formwork	8	4			96	
	Reinforcing						
	Placing	4	6			80	Placing fill on 2nd flr. stl. deck
	Precast						
	Masonry: Brickwork & Stonework						
	Block & Tile	1	1			16	O.E. & Lab unload/stack block
	Metals: Structural			8		64	
	Decks						
	Miscellaneous & Ornamental			3		24	Started today
	Carpentry: Rough						
	Finish						
	Moisture Protection: Waterproofing						
	Insulation						
	Roofing & Siding			7		56	Heavy rain predicted. Roofing sub is rushing to cover as much as possible
	Doors & Windows						
	Glass & Glazing						
	Finishes: Lath, Plaster & Stucco						
	Drywall						
	Tile & Terrazzo						
	Acoustical Ceilings						
	Floor Covering						
	Painting & Wallcovering			1		8	Started today
	Specialties						
	Equipment						
	Furnishings						
	Special Construction						
	Conveying Systems						
	Mechanical: Plumbing			3		24	
	HVAC			5		40	
	Electrical						

Page 1 of 2

Means Forms

EQUIPMENT ON PROJECT	NUMBER	DESCRIPTION OF OPERATION	TOTAL HOURS
Forklift	1	Unloading stock	8
Skiploader	1		8
Flatbed Truck	1		4
Ton Crane	1	Stand By	0

EQUIPMENT RENTAL - ITEM	TIME IN	TIME OUT	SUPPLIER	REMARKS

MATERIAL RECEIVED	QUANTITY	DELIVERY SLIP NO.	SUPPLIER	USE
H.M. Door Frames	20	P 6184	Spec. Supply	Admin. Bldg.
8" Conc. Block	10 m	403	Davis Co.	Admin. Bldg.

CHANGE ORDERS, BACKCHARGES AND/OR EXTRA WORK

VERBAL DISCUSSIONS AND/OR INSTRUCTIONS: Approx. 3 days behind schedule. Will try & make up some time placing concrete.

VISITORS TO SITE

JOB REQUIREMENTS

Page 2 of 2

Identifying the subcontractors on the job helps to confirm the progress schedule. The number of workers is an indication of a subcontractor's efforts to cooperate with schedule requirements. Performance is, of course, the most important criterion—and the basis for the superintendent's decisions on how much to pressure or pay the subcontractors.

Work started and completed today should be included in the information recorded on the Daily Report. When compared to the progress schedule, these records help to identify the actual state of progress relative to the schedule.

The report of equipment on the job site is a means for the office to account for both company owned and rented equipment and to plan for movement from project to yard to project (dispatching).

The report of job progress is a request for the superintendent to state his own interpretation for the benefit of interested, but less informed administrative personnel.

Spaces for remarks provide the superintendent with the opportunity to explain departures from the progress schedule, problems of material delivery, subcontractor performance, inspections, tests and unusual conditions.

Daily reports, once they are started, are not a great burden to the superintendent since the format carries over from day to day with only minor differences.

Diaries

Unlike daily reports, diaries are not usually required of the superintendent. Although encouraged by company management, the use of diaries is left to the initiative and discretion of the superintendent. Diaries can have a valuable influence in the settling of disputes, because they have a certain degree of historic authenticity. If, for instance, a diary records repeated hostile behavior on the part of an inspector, or repeated broken promises on the part of a subcontractor, a court or board of arbitration might be influenced to rule in favor of the superintendent's company. Figure 10.2 is an example page from a superintendent's diary.

Logs

Logs record facts in greater detail than daily reports. Unlike diaries, logs deal with hard facts. They are not read daily by others and, in fact, might never be read by others. The superintendent keeps a log just in case the information might someday be needed. Figure 10.3 is an example of a page from a superintendent's job log.

Photographic Records

The preceding section addressed the value of historical records in the form of daily reports, diaries, and logs. Another form of historical record is the photograph, which furnishes visual authenticity. Photographs should be taken of the conditions that exist before work begins and should concentrate on potential problem areas, such as springs indicating subsurface water, and cracks in the walls of existing adjacent buildings (to rebut charges of liability).

As work proceeds, photographs should be taken at critical or strategic points to prove the proper placement of materials, structural elements, or connections before they are concealed by subsequent work.

Photographs taken by a professional may be specified as a part of the contract requirements. Some optional photos may also be useful to the superintendent for purposes of company history, credit, aesthetic, point of interest, and educational value. Photographs of a project may be gathered together in an album and made part of a growing company reference library.

DIARY   DATE 4/10/88

PROJECT Administration Bldg. & Fire Station, Picacho, CA

Outdoor work was 75% of normal production
due to intermittent showers.

    I highly recommend Topper Roofing Co. for
future jobs.

    After 2 weeks of broken promises, Spencer
Paint Co. finally sent _one_ painter (3 were promised).

    The steel sub had a near-accident at 10:20 a.m.
due to their old, poorly maintained crane - too
small for the job. I notified the steel foreman.
(Copy of written notification is in the file).

                        SUPER   B.B.

Figure 10.2

```
JOB LOG                                    DATE  4/10/88

PROJECT  Admin. Bldg. & Fire Station      Picacho, CA


EXCAVATION     Backfill footing - 66 CY

               Excav. for curbs   - 180 LF

               Fine grade for walks- 1200 SF

FORMWORK      Conc. curbs - 80 LF

              Walk headers - 320 LF

CONCRETE PLACED    Curbs -   6 CY

                   Walks -   8 CY

                   2nd flr. slab- 12 CY

LUMBER PLACED    Misc. blocking - 900 BF


MATERIALS RECEIVED  Casework -   1 Truckload


SUBCONTRACT   Air cond. ductwork started today
   Plumbing fixtures received & stored in yard


REMARKS  Architect made a routine inspection
    and suggested a possible change order to
    reroute the roof downspouts.



                                        B. B. Super
```

Figure 10.3

Watching the Weather

Weather records, mentioned earlier in this chapter, can be of significant historical value. In addition to recording past weather conditions, the superintendent also needs to make value judgments regarding future weather conditions. Planning for the job is based on weather predictions as presented by official bulletins, or by his own best guesses. In the face of uncertainties, the superintendent must decide whether to schedule or cancel certain weather-sensitive work, such as placing and finishing concrete. He must also decide whether to cover and protect work overnight or on weekends, or take a chance on the weather. In emergencies, the superintendent must return to the job site after working hours to take preventive action against weather damage.

The superintendent's weather records could be very important in cases where a project's completion time is delayed because of interfering weather. This kind of documentation might prove invaluable when it comes to obtaining a time extension.

Computers in the Field

The purpose of computers in the business world in general and the construction business in particular is threefold: (1) to speed up paperwork processing and communication, (2) to reduce the incidence of error, and (3) to decrease the human workload. The project superintendent can benefit from all of these improvements, but whether or not he has access to a computer depends upon his company. Sooner or later, most superintendents will face the necessity of using a computer.

The typical, traditional paperwork (as described in previous sections of this book) needs only superficial changes to be adapted to computer processing. Following are examples of the kinds of documentation that can be done on the computer:

Daily Reports can be stored on disk, tape or microfilm. They can be coded and indexed by date. One or more copies may be printed as desired (the original should be kept for its legal value).

Data on subcontractors and subcontracts can be stored on disk or tape for quick and easy access. Copies of the actual subcontracts are kept in the home office file cabinet for in-depth reference. Information such as trades, names of subcontractors, special inclusions and exclusions, total dollar amounts of the contracts, as well as addresses, telephone numbers and names of representatives can be made available at the computer. This method can be a valuable time-saver. Figure 10.4 is an example of data that can be displayed as needed.

Construction Progress Schedule data can be stored on disk or tape and continually or periodically updated. Based on the updated information, the schedule can be redrawn to visually display the current status of job progress. If progress payments are coordinated with the progress schedule, they too can be stored in the computer system, kept updated, accumulated and then printed out for billing.

Submittal expediting chart data may be stored in the computer system and reviewed for status on the terminal, with critical or problem situations highlighted.

Word processing programs can be used to prepare memos to subcontractors, urging them to meet vital deadlines. Word processing can also be used for a variety of other communications, for instance:

- Letters of transmittal, such as the forwarding of submittals to the sponsor, and packages to subs or to the home office.

SUBCONTRACT INFORMATION

CODE

2700

TRADE – Structural Steel

SUB – Ringer Steel Co.

INCLUSIONS – Miscellaneous metal

EXCLUSIONS – Steel decking

CONTRACT – $156,600

ADDRESS – 200 Welder St., Riverside, CA.

PHONE – 545-2200

REP – Paul Ringer

2800

TRADE – Steel Decking

SUB – Dekko

INCLUSIONS – Hoisting

EXCLUSIONS – Paint touchups

CONTRACT – $21,240

ADDRESS – 1000 Sheet Blvd., Redlands, CA.

PHONE – 676-1084

REP – Phil Engels

2900

TRADE – Insulation

SUB – Comfy

INCLUSIONS – Wall & ceiling insulation

EXCLUSIONS – Furring

CONTRACT – $6,100

ADDRESS – 122 Court Lane, Picacho, CA.

PHONE – 297-7077

REP – Jean Comfy

Figure 10.4

- Memos to subcontractors advising them of scheduled meetings.
- Printed documentation of meetings with subs.
- Letters to accompany change orders.
- Work orders.

Cost record systems may be the major category of paperwork processed by a computer. This information can be recorded in the field or by a data communication system and then transferred to a central processing unit in the contractor's home office. Cost items running more than a certain percentage (say 10%) over the original estimate can be highlighted for special attention.

Quality control data can be stored in the computer to create a historical record, a current status review, and a guide for scheduling and expediting tests and inspections.

Data on employees can also be stored for reference in the computer system. Information such as names, addresses, phone numbers, trade clasifications, Social Security numbers and starting and ending dates of employment might be included. Personal comments may also be appended. These records are intended to be confidential, but additional privacy may be achieved by the use of coding as shown in Figure 10.5, or with confidential files accessed with a special entry code or password.

In Figure 10.5, E stands for "excellent"; A stands for "average"; no code indicates either "below average" or "too new to be judged". If the superintendent requires other data such as number of dependents and special skills, this information can also be included.

Change orders can be stored in complete detail for ready reference. Each line item in each change order can be coded for grouping into categories and trades. This is helpful for revising the original cost record system and for updating monthly payment requests.

Progress payments received can be stored on disk or tape as an historical record and can also be used for updating. As payments become due, the sponsor is presented with print-outs, itemizing the costs of work accomplished to date.

When the project is completed, all tapes and disks can be labeled and stored in the company library for future reference.

EMPLOYEE INFORMATION

Skilman, David   **SS#** 000 11 2222

    TRADE – Journeyman Carpenter

    ADDRESS – 2227 Bradford St., Picacho, CA.

    PHONE – 227-2772

    EMPLOYED – 1/18/87 to 4/11/87

    COMMENT – E

Snoozer, Sam   **SS#** 000 11 1111

    TRADE – Common Laborer

    ADDRESS – 207 Starsky Ave., Toolow, CA.

    PHONE – 766-2376

    EMPLOYED – 1/8/87 to 1/12/87

    COMMENT –

Commoner, John   **SS#** 000 11 1112

    TRADE – Cement Finisher

    ADDRESS – 7002 Picacho St., Picacho, CA.

    PHONE – 257-8878

    EMPLOYED – 3/16/87 to 4/20/87

    COMMENT – A

Figure 10.5

Chapter 11

Administrative Meetings

Despite the autonomy of his position, the superintendent need not be an "island". Two types of regular administrative meetings that benefit both the superintendent and the project are company management conferences and subcontractor meetings. Irregular meetings with the sponsor's agents can also be beneficial.

Company Management Meetings

These sessions are usually held at the home office. They are attended by all of the company's project superintendents at selected times when their absence from the job sites matters least. These meetings are opportunities for:
- Constructive criticism
- Reporting of problems
- Exchanging advice, such as methods of construction
- Exchanging information, such as sources of materials and equipment
- Trading of crews and equipment
- Rating the performances of subcontractors
- Updating knowledge of codes, rules, and regulations

Subcontractor Meetings

These conferences are usually conducted in the project field office. Any interested sub may attend; but only those whose trades are currently under way, or soon to start, are specifically invited. The objectives of these meetings are:
- Improvement of teamwork between subs.
- Early discovery of potential problems, such as delays in the delivery of material or manufactured products, or pending trade union strikes.
- Discovery of potential areas of dispute between subcontractors, such as the exact points at which work is separated between trades (jurisdiction).
- Discussion of the progress schedule in general.
- Determination of the quantity of specific items to be furnished by specific dates, and obtaining promises of compliance.

This subject is discussed in greater detail in Chapter 16.

Sponsor Meetings

These sessions may be conducted in the field office or at the sponsor's request. Usually, these meetings center on critical decisions such as:

- Preconstruction questions and answers
- Protection of adjacent properties
- Environmental protection
- Changes in the progress schedule
- Changes in order negotiations
- Test and inspection failures
- Final acceptance of the project

This subject is discussed in greater detail in Chapter 5, "Sponsor Relations".

Chapter 12

The Superintendent's Reference Library

It would not be practical for the superintendent to keep at his fingertips the entire range of reference data relevant to construction. Some information is needed only on rare occasion. Other data is required often enough so that it should be readily available. A portable library is useful, since the superintendent will move it from job to job. A case of shelves makes it possible to move a library without disturbing its arrangement. The case could be made lockable, should the superintendent wish to protect the library from unauthorized researchers.

Following is a list of typical reference data used often enough to rate space in the superintendent's portable library. (Rarely used data can be obtained from off-site sources, such as a public library, or by telephone inquiry.)

1. **A textbook on surveying and laying out techniques** with solutions to various field problems, including a discussion of professional methods, notations, and record keeping.

2. **A first aid handbook** with proper techniques for various emergency situations. This manual should be clearly written and referenced.

3. **A safety code book** covering all local, state and federal laws, rules and regulations. This book should include general guidelines for ensuring safe practices in all aspects of construction. Important examples involve the use of ladders, scaffolding, rigging, trench shoring, floor soffit shoring, heavy equipment, protective devices, electricity, and safety railings.

4. **Construction cost data** such as R. S. Means' current unit prices for various trades. These figures represent averages as reported by thousands of contractors nationwide. This cost information is also useful for estimating and negotiating project change orders. The superintendent can make decisions about construction methods based on comparisons drawn from this cost data.

5. **Building code handbooks** applicable to the area in which the project site is located. With these references available, the superintendent can correctly execute the required details whether or not they are explicitly described in the project drawings and specifications.

6. **A copy of trade union rules and regulations** to guide the superintendent in worker and subcontractor relations. These regulations address such matters as work hours, overtime, and areas of

jurisdiction, as well as subsistence, rest periods, holidays, pay scales, and numerous other benefits and procedures.

7. **A list of construction standards and details** approved and recommended by city, county, and state building and engineering departments. These standards apply to items such as: concrete curbs, sidewalks, driveways, manholes, catchbasins, and barricades. These matters are often not detailed in project drawings; in such cases, the superintendent must turn to accepted standards for guidance.

8. **A handbook of lumber grading and dressing rules.** This resource provides information on stock and premium dimensions, lengths, surface appearances, densities, workabilities, and strengths of different species and classes of lumber and plywood. This data helps the superintendent to select appropriate material for such items as scaffolding, shoring, bracing, concrete formwork, posts, and beams. Information is also provided on the use of beams and girders—both for temporary purposes and permanent structural or framing use.

9. **A concrete formwork design handbook.** This is a source of data on the strength, dimension, and length of members and the spacings of ties. Such a handbook can also help the superintendent in choosing methods and techniques for special formwork, such as panelizing, partial removals (during curing), shoring, and reuses of materials.

10. **A handbook for concrete and concrete placement and finishing.** This volume should provide data on mix designs suitable for various strength requirements, weather conditions, workabilities, and curing rates. Additional information and instructions should include the methods of concrete placement under various conditions, such as limited working room, height above ground, under water, in tall or thin forms, and so forth.

11. **A concrete accessories catalog.** This catalog should provide data useful in the selection of various materials and equipment. Examples are form ties, scaffolding supports, braces, clamps, tube forms, column clamps, form liners, screed chairs, curing compounds, concrete hardeners, joint sealants, bond breakers, placing equipment, waterstops, and precast concrete inserts.

12. **A rough hardware catalog.** Included in this catalog should be data for the selection of various material and equipment such as anchor bolts, machine bolts, beam hangers, framing anchors, post caps and bases, strap ties, joist bridging, nails, saws, hammers, and other small tools.

13. **A structural steel manual.** This manual should provide data on various stock shapes, sizes, weights, and the strengths of various metals. The superintendent needs this information to select and order metal for temporary or permanent construction work. Items that should be covered include bracing, beams, connectors, plates, columns, lintels, door bucks, and angle frames.

14. **Architectural and engineering textbooks.** These are needed to provide a wide variety of data on construction materials and methods, as well as architectural, mechanical and electrical trades, and mathematical tables and formulas.

15. **An equipment source and rental rates catalog.** Information on various equipment, makes, models, sizes, where to obtain, average rental rates and production rates should be available. This information helps the superintendent to choose the most suitable and economical equipment in the quickest possible time.

16. **English dictionary.** This is a standard office item, but the superintendent should have a copy as part of his personal library.

17. **Construction Dictionary.** Over the years, the superintendent encounters unfamiliar terms that describe new or old, familiar and unfamiliar materials or procedures. A good reference such as *Means Illustrated Construction Dictionary* provides definitions and helps to preserve the superintendent's reputation for knowledge and experience.

Chapter 13

Cost Controls

It is possible that small, but worthwhile savings may be made in materials by minimizing waste, reusing (as in formwork), and carefully measuring and installing good security provisions (to deter theft and vandalism). Discounts are another potential source of savings. Figure 13.1 is an analysis of "loose", as opposed to "tight" materials management.

Total cost and total savings are of primary interest. In some cases, a greater net saving may be realized by deliberately using a larger quantity of material than estimated and budgeted in order to employ less labor. The following charts compare estimated to actual costs for concrete footings.

Estimated

	Quantity	Material		Labor		Total
		u.c.	mat'l	u.c.	labor	
Side forming	1,600 sf	.27	$432	1.35	$2,160.00	2,592
Concrete	45 cy	50.90	2,291	7.23	325.00	2,616
Ex. & backfill	30 cy		0	5.60	168.00	168
Total			$2,723		$2,653.00	$5,376

Actual

	Quantity	Material		Labor		Total
Top forming	800 lf	.37	296	1.56	1,248	1,544
Concrete	60 cy	50.90	3,054	7.23	434	3,488
Ex. & backfill	15 cy	0	0	8.00	120	120
Total			$3,350		$1,802	$5,152

By excavating footing trenches "neat" and eliminating side forms, but retaining minimum layout and top pour strips (Figure 13.2), the labor cost is reduced, even though more concrete material is used.

Methods used to control labor and equipment production deserve more of the superintendent's attention than methods designed to save on materials. Potential materials savings or losses are minor compared to labor and equipment costs. The previous outline for materials savings is simple economics, but controlling labor and equipment costs calls upon the superintendent to use a greater degree of expertise.

The term "savings" refers to the difference when an actual cost is subtracted from the budgeted amount. The term "loss" refers to the difference when the budgeted amount must be subtracted from the actual cost. Chapter 14 addresses the issue of time controls, time being the real key to controlling costs. Regarding both labor and equipment,

hourly rate × time = cost.

Material Wastes	**Material Savings**
• Purchase from the estimator's quantity survey.	Superintendent carefully draws up his own material list.
• Purchase from the first or second supplier quotation.	Purchase the best of three or four quotations.
• Purchase without considering possible substitutes.	Consider substitute brands acceptable to the sponsor.
• Purchase material for a complete job, without considering reuses.	Purchase a fraction of the total material needed, and calculate for reuses.
• Dispose of used materials at the dump.	Sell used materials at salvage value.
• Use a careless approach in cutting and ordering.	Minimize waste by careful planning, ordering, and cutting.
• Neglect security and weather protection measures.	Provide protection, covering and locked storage.

Figure 13.1

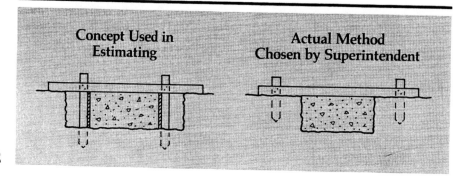

Concept Used in Estimating	Actual Method Chosen by Superintendent

Figure 13.2

There is no cost control in this statement. To have cost control, we must introduce two more terms; output and unit cost, thus:

$$\frac{\text{hourly rate}}{\text{output}} = \text{unit cost}$$

Example:

$$\frac{\$18/hr}{9 \text{ items/hr}} = \$2/\text{item}$$

Unit cost is not of primary importance per se, but it helps to evaluate output or production and leads to the final, most important figure: the total cost. Output and time have an inverse relationship, as shown in Figure 13.3.

The time required to do an item of work is a function of output; as output increases, total time decreases, and vice versa. In general, the superintendent need only think about maximizing production and he can almost ignore cost, since minimum cost is the automatic result. To compare actual to estimated production, he must perform simple math. For example:

Given Budget

	Quant	Unit Cost	Budget	Hours	Production
Trench excav.—mach.	133 cy	$4.06	$540	18	7.40 cy/hr
Trench excav.—hand	25 cy	$16.10	$403	25	1.00 cy/hr
Form slab edges	900 lf	$1.25	$1,125	12	75.00 lf/hr
Form footing sides	1,600 sfca	$1.35	$2,160	27	60.00 sfca/hr
Place concrete	45 cy	$7.23	$325	3	15.00 cy/hr

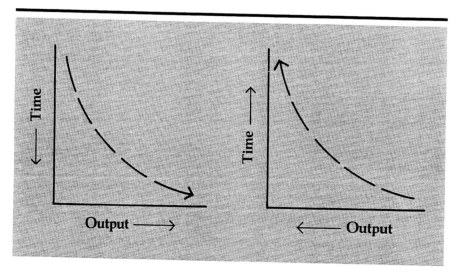

Figure 13.3

With the information under the Production column, as in Figure 13.4, the superintendent can observe the actual performance relative to the hourly and daily budget. He may increase or decrease the sizes of crews, change units of equipment, devise better construction methods, or take other steps to counteract slipping production rates.

The importance of this method is that it enables the superintendent to know at all times the status of actual production while there is still time to take remedial action.

The following is a list of suggested ways to save on labor:
- Perform extensive advance planning, detailing and layout work so that labor will be efficient and free from delays, stops, and complications.
- Employ optimum size crews. Too many or too few workers may be uneconomical.
- Plan the work and schedule the crews so that work is done in full or half day allotments with no remaining unproductive hours.
- Plan miscellaneous work items to employ those workers who finish other jobs early in the day, or are awaiting material or equipment delivery.
- Consider using various tools or equipment to aid or replace labor.
- Consider offering bonus incentives to workmen to increase production.
- Keep in mind and hold back on work items that can be accomplished during inclement weather.

The subject of savings in labor costs is discussed further in Chapter 15, "Productivity and the Manning Chart".

The list below suggests ways to save on equipment:
- Use the optimum size of equipment. A unit that is too large or small for the task will not be economical. A job which takes more than a day or two might warrant the expense of bringing in the proper equipment rather than using a unit only because it happens to be already present on the site.
- Schedule the work so that multiple uses can be made of costly rental equipment, such as a large crane.
- Consider purchasing, rather than renting equipment when a project is large enough to make the purchase economically feasible. For example, a water pump may be required for 300 hours at a rental rate of $8/hour. The total estimated cost is $2,400; whereas, a new pump may be purchased for $3,000 and a depreciation rate charged to the project.

The Cost Record System

The cost record system contributes to cost control, as it keeps the superintendent aware of the budget limitations of each separate item during the competitive bidding stage. By regularly plotting individual expenditures, the superintendent is able to spot cost problems and apply special efforts to minimize them. It is understood that the superintendent steadily restrains all cost items as a normal practice, but special efforts may be necessary for items which threaten to exceed the budget by large amounts. Such preventive measures might involve:
- Devising better, more efficient methods of construction.
- Attempting to increase production by changing workers, sizes of crews, and/or types of equipment.
- Increasing the proportion of prefabrication.

Figure 13.4 is an excerpt from a sample cost record system. It contains the original cost breakdown for bidding purposes. Some items (*) have been completed, some are in progress, and others have not yet been started.

Administration Building & Fire Station, Picacho, CA.
ESTIMATED COST

PAY ITEM	COST CODE	WORK DESCRIPTION	QUANTITY	LABOR U/C	MAT'L U/C	SUPER'S INFORMATION TOTAL HOURS	PRO-DUCTION
100	A100	GENERAL CONDITIONS					
	A100-2	Layout of structures	48 hr	52.61	3.00	48	150 LF/Hr. *
	A100-3	Field office	11 mo	------	85.00	—	—
	A100-4	Storage shed	11 mo	------	75.00	—	—
	A100-5	Toilets (2 ea)	11 mo	------	95.00	—	—
	A100-6	Telephone	11 mo	------	150.00	—	—
	A100-7a	Water hookup & meter	ls	------	ls	—	— *
	A100-7b	Water monthly charges	ls	------	60.00	—	—
	A100-8a	Elec. service & transf.	ls	------	ls	—	— *
	A100-8b	Elec. monthly charge	11 mo	------	70.00	—	—
	A100-9	Security yard fence	200 lf	------	2.00	—	—
	A100-10	Project sign	ls	ls	ls	—	—
	A100-11	Temp. weather closures	ls	ls	ls	—	—
	A100-12	Office supplies	11 mo	------	35.00	—	—
	A100-13	Small tools	11 mo	------	200.00	—	—
	A100-14	Misc. rental equipment	11 mo	------	400.00	—	—
	A100-15	Oil, fuel, repair, service	11 mo	------	150.00	—	—
	A100-16a	Cleanup-progressive	11 mo	1,200.00	100.00	—	—
	A100-16b	Cleanup-final	240 hr	15.10	5.00	—	—
	A100-17	Not used	------	------	------	—	—
	A100-18	Superintendent	11 mo	3,220.00	------	—	—
	A100-19	Subsistence	209 day	------	20.00	—	—
	A100-20	Barricades (pedestrian)	700 lf	6.00	6.00	247	3 LF/Hr.
	A100-21	Night lights	11 mo	------	60.00	—	—
	A100-22	Not used	------	------	------	—	—
	A100-23	Safety railings	280 lf	1.50	1.00	23	12 LF/Hr.
122	A130	EARTH SHORING	260 lf	10.67	15.69	163	2 LF/Hr. *
123	A131	EARTHWORK (GRADING)					
	A131-1a	Exc down to buried mat'l	17,000 cy	------	.80	170	100 CY/Hr.*
	A131-1b	Haul away A131-1A mat'l	17,000 cy	------	2.05	283	60 CY/Hr.*
	A131-2a	Exc & load debris	20,000 cy	------	1.13	267	75 CY/Hr.*
	A131-2b	Haul away A131-2a mat'l	20,000 cy	------	2.25	400	50 CY/Hr.*
	A131-3a	Breakup concrete	5,000 cy	------	7.68	320	16 CY/Hr.*
	A131-3b	Haul away A131-3a mat'l	5,000 cy	------	3.92	167	30 CY/Hr.*
	A131-4a	Purchase import mat'l	34,500 cy	------	2.95	575	60 CY/Hr.*
	A131-4b	Spread/compact	34,500 cy	------	.90	207	167 CY/Hr.*
	A131-5	Fine grade overall	120,000 sf	------	.03	80	1500 SF/Hr.*
	A131-6	Erosion protection	9,000 sf	------	.15	23	391 SF/Hr.*
	A131-7	Dump fees	2,200 lds	------	15.00	—	— *
	A131-8	Equip. move on/off	80 hrs	------	50.00	—	— *
	A131-9	Supervision	5 wks	------	1,176.00	—	— *

*line item completed

Figure 13.4

Administration Building & Fire Station, Picacho, CA.

PAY ITEM	COST CODE	WORK DESCRIPTION	QUANTITY	LABOR U/C	MAT'L U/C	TOTAL HOURS	PRO- DUCTION
		ESTIMATED COST				SUPER'S INFORMATION	
132	A150	PARKING STRIPING	3,000 lf	.21	.10	15	200 LF/Hr.
138	A151	CONCRETE WORK & STRUC. EXC.					
	A151-1a	Footings, layout	1,200 lf	.88	.10	62	19 LF/Hr.*
	A151-1b	Machine exc	320 cy	-----	8.00	48	7 CY/Hr.*
	A151-1c	Hand trim trenches	2,800 sf	.40	-----	75	37 SF/Hr.*
	A151-1d	Forming	1,200 lf	2.95	.90	200	6 LF/Hr.*
	A151-1e	Concrete & placing	220 cy	8.26	65.00	113	2 CY/Hr.*
	A151-1f	Concrete cure ftg tops	1,200 lf	.35	.035	24	50 LF/Hr.*
	A151-1g	Disposal excess dirt	220 cy	-----	5.20	6	37 CY/Hr.*
	A151-1h	Backfill & tamp	100 cy	9.00	3.00	56	2 CY/Hr.*
	A151-2a	Flr slab on grd-layout	6,400 sf	.12	.03	45	142 SF/Hr.*
	A151-2b	Fine grade for gravel	6,400 sf	.10	.05	40	160 SF/Hr.*
	A151-2c	4" gravel base course	95 cy	9.37	21.58	56	2 CY/Hr.*
	A151-2d	6 mil poly membrane	7,000 sf	.07	.05	29	241 SF/Hr.*
	A151-2e	2" sand & fine grade	48 cy	12.37	24.80	37	2 CY/Hr.*
	A151-2f	Form slab edges	800 lf	3.00	1.20	141	6 LF/Hr.*
	A151-2g	Form depressions	164 lf	2.90	.60	28	6 LF/Hr.*
	A151-2h	Anchor bolts for sills	200 ea	3.25	.80	38	5 Ea./Hr.*
	A151-2i	Set screeds	6,400 sf	.10	.04	38	168 SF/Hr.*
	A151-2j	Concrete & placing	90 cy	8.26	65.00	46	2 CY/Hr.*
	A151-2k	Concrete pumping (1/2 of slab)	45 cy	-----	12.00	8	6 CY/Hr.*
	A151-2l	Finish & cure	6,400 sf	.30	.03	113	57 SF/Hr.*
	A151-2m	Sealer per fin. sched.	4,000 sf	.07	.07	16	250 SF/Hr.*
	A151-3a	Steps on grd(tread lgth)	22 lf	25.00	2.50	33	1 LF/Hr.*
	A151-4a	Steps to 2nd flr-forming	640 sf	5.90	2.80	222	3 SF/Hr.*
	A151-4b	Concrete & pumping	6 cy	22.00	80.00	8	1 CY/Hr.*
	A151-4c	Finish & cure	240 sf	1.20	.20	17	14 SF/Hr.*
	A151-5a	Fill on stl deck- screeds	2,200 sf	.12	.06	17	129 SF/Hr.*
	A151-5b	Concrete & pumping	22 cy	16.00	80.00	22	1 CY/Hr.*
	A151-5c	Finish & cure	2,200 sf	.30	.03	39	56 SF/Hr.*
	A151-6a	Curbs on slabs	68 lf	2.75	.75	11	6 LF/Hr.*
	A151-7a	Locker bases	50 lf	4.00	2.00	12	4 LF/Hr.
	A151-8a	Equipment slabs	60 sf	3.50	2.00	12	5 SF/Hr.
	A151-9a	Splash blocks	6 ea	32.00	32.00	11	1/2 Ea./Hr.
	A151-10a	Catch basins	4 ea	480.00	255.00	113	
139	A180	CARPENTRY					
	A180-1a	Doors, Hollow metal	6 ea	39.88	-----	13	1/2 Ea./Hr.
	A180-1b	Doors, wood	18 ea	39.88	-----	40	1/2 Ea./Hr.
	A180-2a	Door frames, hol met	24 ea	29.91	-----	40	2/3 Ea./Hr.
	A180-3a	Hardware, closers	12 ea	19.94	-----	13	1 Ea./Hr.
	A180-3b	exit devises	3 ea	39.88	-----	7	1/2 Ea./Hr.
	A180-3c	thresholds	6 ea	29.91	-----	10	2/3 Ea./Hr.
	A180-3d	seals	6 ea	39.88	-----	13	1/2 Ea./Hr.

*line item completed

Figure 13.4 continued

Administration Building & Fire Station, Picacho, CA.

PAY ITEM	COST CODE	WORK DESCRIPTION	QUANTITY	ESTIMATED COST		SUPER'S INFORMATION	
				LABOR U/C	MAT'L U/C	TOTAL HOURS	PRODUCTION
139	A180	CARPENTRY (CONT'D.)					
	A180-4a	Window frames (arched)	9 ea	199.70	-------	105	12 Hrs. Ea.
	A180-4b	Window frames (rectangular)	25 ea	159.00	-------	220	9 Hrs. Ea.
	A180-5a	Single storage shelving	38 lf	3.50	-------	7	5 LF/Hr.
	A180-6a	Receptionist counter	24 lf	7.00	-------	9	3 LF/Hr.
	A180-7a	Base cabinets	30 lf	7.00	-------	12	3 LF/Hr.
	A180-8a	Janitor shelf	12 lf	3.50	-------	3	4 LF/Hr.
	A180-9a	Firemen's bunks	2 ea	80.00	-------	9	4½ Hrs. Ea.
	A180-10a	Kitchen Base Cabinets	10 lf	8.00	-------	5	2 LF/Hr.
	A180-10b	Kitchen wall cabinet	30 sf	4.50	-------	8	4 SF/Hr.
	A180-11a	Work bench	12 lf	6.00	-------	4	3 LF/Hr.
	A180-12a	Wall shelving units	160 sf	2.00	-------	18	9 SF/Hr.
	A180-13a	Closet shelves	36 lf	2.50	-------	5	7 LF/Hr.
	A180-13b	Closet poles	36 lf	1.50	-------	3	12 LF/Hr.
	A180-14a	Shower bench	8 lf	8.00	-------	4	2 LF/Hr.
	A180-15a	Roof nailers, 2x4;2x6	400 lf	.70	.70	16	25 LF/Hr.
	A180-16a	Roof fascia, 2x12	480 lf	1.20	1.60	32	15 LF/Hr.
	A180-17a	Misc. rough hardware	ls	-------	ls	—	—
	A180-18a	Install toilet accessories	24 hrs	19.94	-------	24	—
139	A181	INSTALL MISC. METAL					
	A181-1a	Fireman's pole	1 ea	80.00	-------	4	4 Hrs Ea.
	A181-2a	Ladders	2 ea	28.50	-------	3	1½ Hrs. Ea.
	A181-3a	Channel door frames	4 ea	40.00	-------	9	2 Hrs. Ea.*
	A181-4a	Door corner guards	8 ea	19.00	-------	8	1 Hr. Ea.
	A181-5a	Lintels over doors	4 ea	60.00	-------	13	3 Hrs. Ea.*
	A181-6a	Pipe guard posts	12 ea	9.50	-------	6	2 Ea./Hr.
	A181-7a	Detention bars	2 ea	160.00	-------	9	4½ Hrs. Ea.
	A181-8a	Stair railings	80 lf	3.00	-------	13	6 LF/Hr.
141	A182	PURCHASE WOOD DOORS	18 ea	-------	70.00	—	— *
146	A183	WALL LOUVERS	8 ea	27.21	80.00	12	1½ Hr/Ea.
166	A184	WINDOW & FIXTURE CLEANING	ls	ls	-------	88	
167	A185	SITE CONCRETE WORK					
	A185-10a	Sidewalks & apron	6,150 sf	-------	1.90	—	—
	A185-11a	Curbs, gutter type	600 lf	-------	9.25	—	—
	A185-12a	Driveways	1,000 sf	-------	2.75	—	—

*line item completed

Figure 13.4 continued

The portion of the Master Cost Record (a complete version appears in *Bidding for the General Contractor*, Paul Cook, R.S. Means 1985) reproduced here contains two columns for the superintendent's information: the approximate total man-hours and the production rates. These unit rates represent the average output of one worker or of one piece of equipment. Using these figures as a reference, the superintendent can easily judge whether actual production is running over or under the budget.

Time and Cost Controls

Time and cost records can be combined by using the progress schedule (Figure 13.5), and the cost record (Figure 13.4). For example, the excavation work is scheduled to be completed in 25 days; but how can the superintendent be certain that the schedule will be met? Even if the work is sublet, he needs to know the true state of progress, rather than relying on the subcontractor's assurances. One method is using a bar chart with scheduled days and production rates, as in Figure 13.6.

The scheduling steps are explained as follows:

Each earthwork operation is given the maximum possible number of days within the 25 day limitation. These time limits, combined with a lack of proper planning, could lead to an equipment traffic jam that jeopardizes efficiency. The goal is to employ machinery at optimum capacity, rather than attempting to increase production simply by increasing the number of units.

In Figure 13.4, A131-1a is limited to seven days for the excavation of 17,000 cy of old fill material at the average rate of 100 cy per hour per excavator. Three excavators will be required to meet the schedule. The same formula applies to each of the other operations.

A131-4a, importation of fill material, is of particular concern. The superintendent will need assurance from the supplier that 34,500 cy will be delivered at the rate of 480 cy per hour continuously for nine consecutive days.

The time control portion of this system is obvious; the cost control portion is not so clear. A good general rule is: if the actual time equals or betters that scheduled, the cost should be within the budget.

Here is one more example of time/cost control. The progress schedule provides 10 days (two 5 day weeks) to place 1,200 lf of footings and to pour 6,400 sf of floor slab. This seems like an impossibly short time period until we consult a bar chart with the computed average production rates, Figure 13.7. By summing up the total hours required to complete these two tasks and dividing by the estimated 80 available working hours, we immediately realize that

$$\frac{1{,}219 \text{ (man-hours for concrete work as based on the estimate)}}{80 \text{ (available working hours)}}$$

= 15 workers will be required.

When the bar chart data are extended and summed up, the number of workers is found to be 14.8.

The number of workers (15) is only an average figure for the 10 day period. In reality, there will probably be a larger number, averaging fewer than 10 days each. The superintendent will try to minimize the number of workers to be hired by transferring them from one work item, as it is finished, to another.

Secondary benefits from this manning chart are: (1) improved predictability and availability of the numbers and classifications of workers, and (2) feedback to the company estimator for educational purposes.

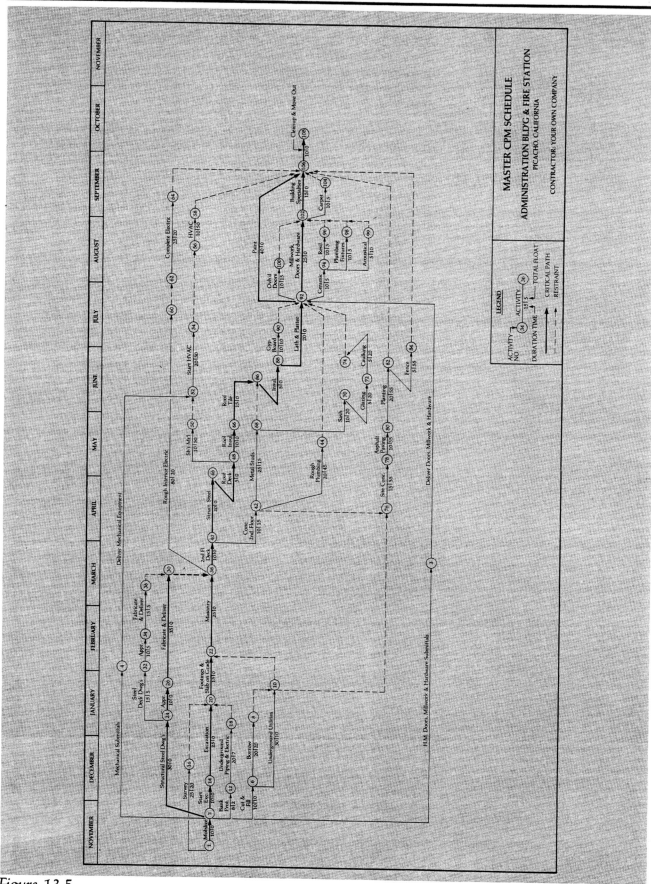

Figure 13.5

EXCAVATION WORK
PRODUCTION COST CONTROL

CODE	WORK ITEM	QUANTITY	25 DAYS TOTAL
A131-1a	Exc. Down to Buried Material	17,000 C.Y.	3 Units @ 100 C.Y./Hr.
A131-1b	Haul Away A131-1a Material	17,000 C.Y.	5 Units @ 60 C.Y./Hr.
A131-2a	Excavate & Load Debris	20,000 C.Y.	4 Units @ 75 C.Y./Hr.
A131-2b	Haul Away A131-2a Material	20,000 C.Y.	6 Units @ 50 C.Y./Hr.
A131-3a	Break Up Concrete	5,000 C.Y.	7 Units @ 16 C.Y./Hr.
A131-3b	Haul Away A131-3a Material	5,000 C.Y.	4 Units @ 30 C.Y./Hr.
A131-4a	Purchase Import Material	34,500 C.Y.	8 Units @ 60 C.Y./Hr.
A131-4b	Spread/Compact Import Material	34,500 C.Y.	3 Units @ 150 C.Y./Hr.
A131-5	Fine Grade Overall	120,000 S.F.	2 Units @ 1500 S.F./Hr.
A131-6	Erosion Protection	9,000 S.F.	1 Unit @ 280 LF/HR
A131-7	Dump Fees	2,200 LDS.	Continuous
A131-8	Equipment Move On/Off	80 Hrs.	80 Hrs.
A131-9	Supervision	5 Wks.	Continuous

Figure 13.6

PRODUCTION RATES SCHEDULE

10 DAYS TOTAL

CODE	WORK ITEM	(bar chart)	WORKERS x DAYS	TOTAL DAYS
A151-1a	Footings - Layout	3 Men	3 x 3 =	9
	Machine Excavation	2 Units	-----	-----
	Hand Excavation	3 Men	3 x 3 =	9
	Formwork	6 Men	6 x 4 =	24
	Concrete	5 Men	5 x 3 =	15
	Curing	1 Man	1 x 3 =	3
	Disposal	1 Unit	-----	-----
	Backfill	3 Men	3 x 2 =	6
A151-2a	Floor Slab - Layout	3 Men	3 x 3 =	9
	Fine Grade	2 Men	2 x 3 =	6
	4" Gravel Base	3 Men	3 x 2 =	6
	Membrane	3 Men	3 x 1 =	3
	2" Sand Base	3 Men	3 x 1 =	3
	Form Edges	4 Men	4 x 4 =	16
	Depressions	2 Men	2 x 2 =	4
	Anchor Bolts	2 Men	2 x 2 =	4
	Set Screeds	2 Men	2 x 2 =	4
	Concrete	5 Men	5 x 2 =	10
	Pumping	1 Unit	-----	-----
	Finish/Cure	5 Men	5 x 2 =	10
	Sealer	2 Men	2 x 1 =	2
			Total Man Days =	143
			143 ÷ 80/8 = Total Number of Workers =	14.3

Figure 13.7

Chapter 14
Time Controls

The most important achievement of the superintendent is completing the project in the least possible time. No one who has a financial interest will argue with speedy completion (assuming the standard of materials and workmanship remains acceptable to the sponsor). If early completion is likely, the superintendent should inform the main office. In this way, arrangements can be made for renting or otherwise using the building as soon as it is available.

Materials should be accurately taken off by the estimator and competitively priced, leaving little scope for cost underruns. On the other hand, labor, equipment, methods of construction and the management of subcontractors are potentially time and money saving areas. While the estimator usually assumes an average production rate, a resourceful superintendent strives to do better through small but numerous time savings. These savings can add up to an impressive early project completion. There are exceptions, such as when the superintendent takes charge of projects with very low estimated budgets. In these cases, it is a challenge just to complete the project on schedule, and very unlikely that it will be finished early.

Figure 14.1 is an example of a tight progress schedule. The estimator judged the ideal completion time at 12 months, but in the follow-up, the expeditor designed the progress schedule for an optimistic 11 months. The superintendent will be congratulated if he actually finishes in 11 months. He takes the following actions:

1. He first notifies every subcontractor and supplier of the scheduled starting and completion dates. This is most effectively done by giving subcontractors copies of the progress schedule with their individual trades inked in red.

2. The superintendent then contacts each subcontractor by telephone or in person and discusses the practicality of the schedule, making adjustments if necessary, and receiving direct agreement and cooperation. Sometimes promises are quite easily made, but not so conscientiously kept. The failure of a subcontractor to perform as promised can be cause for the cancellation of the contract. Having a record that the subcontractor knew the schedule and agreed to it can be a decisive factor in the superintendent's favor should there be litigation (see Chapter 10—"Record Keeping").

Figure 14.2 is an example of a typical letter to subcontractors putting the matter on record. The superintendent's files should contain a folder for each subcontractor with copies of all correspondence and memos. In addition, telephone and personal oral communications (as well as the names of witnesses, if any) should be recorded in the superintendent's daily diary.

3. Next, the superintendent expedites the submittals for approval by the sponsor's architect/engineer. The submittals involve shop drawings,

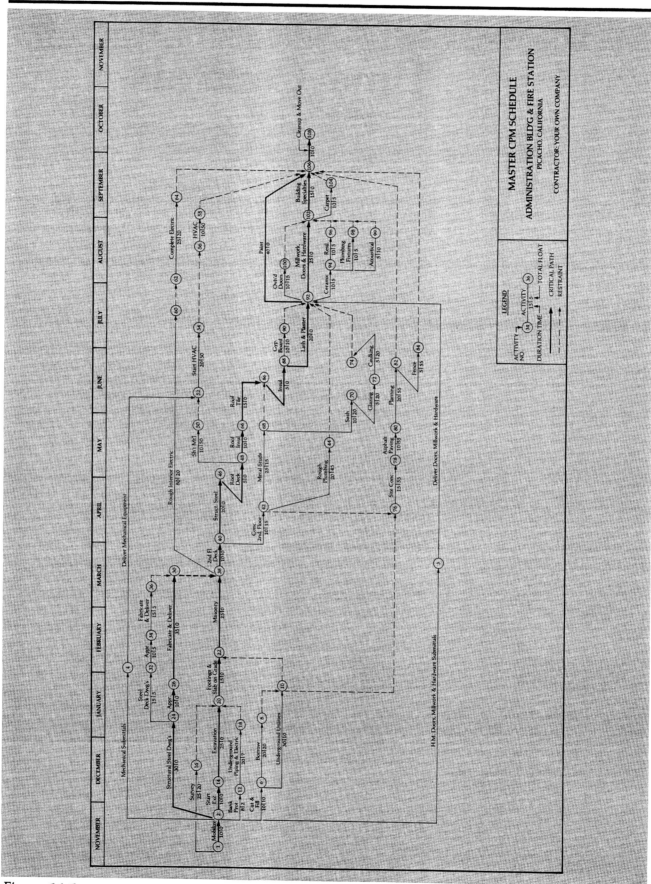

Figure 14.1

  Date

Blazers
770 Interloper Place
Toolow, CA 92304

ATT: Mr. Ted Blazer, Manager

RE: Administration Building & Fire Station

Dear Mr. Blazer:

     Enclosed is one copy of the progress schedule which
has been approved by the sponsor and which confirms previous
oral agreement between you and me, made on November 5, 1986.
Your work, which is red-lined on the schedule, will start
approximately September 9, 1987 and be completed approximately
September 30, 1987. Expect a call from us to notify you of the
exact date and to confirm your readiness to start.

                              Very truly yours,

                              Your Own Construction Company

                              B. Buttress
                              Superintendent

Encl: Progress schedule

Figure 14.2

material samples, catalog data and certifications from subcontractors, with priority given to those needed earliest. Figure 14.1 realistically provides time spans for this process. A separate form is helpful for organizing this important time control activity. The superintendent need not personally carry out all of the time consuming desk work of expediting; most of it can be done by the clerk while the superintendent is attending to site construction activities. The superintendent should, however, be aware of the progress of submittals, and exert pressure upon the slow and late submitters. (Chapter 16 presents measures for dealing with subcontractors who default in furnishing submittal data.)

Figure 14.3 is an example of an expediting chart for the use of the superintendent, the clerk and others. It serves as a guide, and records the steps that need to be taken, as well as the progress that has been made. Priority is given to the trades which (1) start immediately, or (2) require a lot of lead time for shop drawings, detailing, or fabricating. It would be unreasonable to press too hard at this stage for data from a trade which is not scheduled to start work in the near future. For this reason, the chart is divided into three sections: earliest, intermediate, and later.

Dates taken from the progress schedule (Figure 14.1) for the on-site start of individual trades are scheduled under columns 3, 5, and 7. Working backward from those, other critical dates are established. Allowances are made for the following activities and eventualities:

- Mobilization of the subcontractor(s)
- The possibility of disapprovals and the need for time to resubmit.
- Review of the submittals for compliance by the superintendent, expeditor, or sponsor's representative.
- The possibility of technical or other problems which cause subcontractors to delay finishing submittal data.

Chances are, all subcontractors will not respond promptly to a request for submittal data. The superintendent must keep a close watch, and when the lead time runs out make a second, urgent request followed up with a memo and a copy for the record. If the case is sufficiently urgent, the memo should warn the subcontractor of his contractual liability for damages due to unjustified delays. The superintendent might also mention the fact that prompt processing of the subcontractor's request for payment is dependent on quick action regarding requested submittal information. In column 2b, three trades have given the superintendent cause for making urgent second requests. Figure 14.4 is an example of a follow-up memo.

In cases which involve complex submittals, the probability of some disapprovals is great; and because of the additional time required for corrections, resubmittals and second (or third) checkings by the sponsor A&E, the superintendent should provide generous lead time in the schedule.

In column 2e of Figure 14.3, the submittals of three trades met with disapproval. Consequently, the masonry work became a critical matter. Structural steel work was affected because of an infringement on the time needed for plant fabrication. Painting was not so drastically affected because of its generous lead time.

When time is extremely tight, it is tempting for the superintendent to rush submittals to the sponsor without first checking them; but a brief check might detect noncompliances which can be corrected quickly, saving days of delay. As a rule, the superintendent is not equipped to recognize all of the fine points of submittal compliance—nor is this necessary. In Figure 14.3, only one or two days are allowed between receiving data from subcontractors and submitting it to the sponsor's architect and engineer.

Figure 14.3

Means Forms

JOB
PROGRESS REPORT

PROJECT Admin. Bldg. & Fire Station
LOCATION Peacho, CA
PREPARED BY

SUBMITTAL EXPEDITING RECORD

SHEET 1 of 2
JOB NO 110
YEARS 1986 – 1987

Trade	Notified Sub of Prog. Sched.	Earliest Submittals							Intermediate					Later Submittals				
		2a 1st Req. of Sub	2b 2nd Req. of Sub	2c Received from Sub	2d 1st Submtl to Sponsor	2e Dis-Approved	2f Approved	3 SubStart work	4a Requested of Sub	4b Received from Sub	4c Submitted to Sponsor	4d Approved	4e Sub Notified	5 SubStart work	6a Requested of Sub	6b Received from Sub	6c Submitted to Sponsor	6d Approved	6e Sub Notified	SubStart work
Surveying	ns							11/5/86											
Demolition	ns 11/5/86							11/7/86											
Earthwork	ns 11/5/86							11/9/86											
Reinf. Steel	* 11/6/86	11/20/86	11/27/86	12/5/86	12/10/86	12/30/86	12/11/86	12/31/86	2/1/87	2/8/87	2/18/87		2/18/87	3/1/87						9/24/87
Masonry	* 11/8/86	11/24/86		12/15/86	12/15/86		18/24/86	1/15/87	3/5/87	3/12/87	3/26/87	4/12/85	4/4/87	3/1/87						9/26/87
Struc./Misc. Stl	11/8/86			1/13/87	1/15/87	1/30/87	1/20/87		4/1/87	4/3/87	4/3/87	4/13/87	4/4/87	5/1/87						9/26/87
Steel Decking	11/8/86			11/20/86	11/20/86		1/21/87	3/1/87					4/14/87	5/1/87						9/4/87
Framing for Gyp. Bd.	11/11/86			11/26/86	11/26/86		1/22/87		1/28/87		5/18/87			7/24/87						10/1/87
Millwork																			
Roofing	11/4/86			11/10/86	11/10/86	12/31/86		11/14/86	1/5/87											
Painting	11/10/86		2/16/87	2/2/87	2/2/87	3/4/87	3/4/87	5/1/87	4/5/87	4/9/87	4/28/87		6/1/87	6/1/87						
Sheet Metal	11/10/86			12/1/87	12/11/87			5/1/87	4/6/87	4/21/87	5/11/87	5/9/87		6/11/87						
Insulation									4/21/87	4/20/87	4/22/87			6/1/87						
Gypsum Board									4/6/87	4/8/87	4/10/87			6/1/87						
Lath & Plaster									4/12/87										
Acoustic Tile									4/6/87										
Hardware	11/8/86	12/3/86		10/25/86	12/8/86		1/5/87	1/5/87	4/20/87	4/3/87	4/17/87		1/25/87	7/25/87						
H.M. Door Frms. *	11/8/86								5/1/87	5/5/87	5/7/87			7/25/87						
Windows																			
Ceiling Doors																			9/24/87
Glass																			9/26/87
Ceramic Tile																			9/26/87
Resil. Flooring																			9/4/87
Specialties																			9/11/87
A.C. Paving																			10/1/87
Caulking	ns																		
Fencing																			
Mech. Underg'd *	11/5/86	11/9/86		11/19/86	11/9/86	12/15/86	12/1/86	12/15/86	2/19/87	1/18/87				Cont						Cont
- Rough	11/5/86	11/5/86		11/24/86	11/2/86	12/15/86	12/1/86	12/15/86											
- Finish																			
Elec. Underg'd *	11/5/86	11/6/86		11/10/86	11/10/86		1/7/87	12/19/86	5/18/87	1/9/87				Cont						Cont
- Rough	11/20/86			12/15/86	12/13/86														
- Finish																			

DATE
YEAR

LEGEND: — = Not Applicable; * = Urgent; Cont = Ongoing work (no fixed date); ns = No Submittal Required
⊠ = Awaiting Requested Submittals or Approvals

  Date

        Betbar
        Ten Bender Lane
        Steelgap, Nevada 89111

        ATT: Mr. Neil Bender

        RE: Administration Building & Fire Station
             Picacho, California

        Dear Mr. Bender:

            This is to confirm my phone call of this date urging
        that you deliver to us no later than 12/5/86 all footing
        and foundation shop drawings. This submittal deadline will
        provide the A & E a necessary minimum of 10 days for checking
        and approval; then you will have only 15 days left to fabricate
        and deliver the steel to the site ready for start of placing on
        12/31/86.

            Please refer to clauses D and F in your subcontract
        regarding liability for damages caused by delays.

                                        Very truly yours,

                                        Your Own Construction Company

                                        B. Buttress
                                        Superintendent

        Copy: Main office
               File

Figure 14.

Whenever he wishes, the superintendent may draw a heavy line to define the current status. In Figure 14.3, everything to the left of the line is completed and everything to the right is future action. This system of submittals can be further expedited using a computer.

4. At this point, the superintendent should be holding periodic meetings with groups of subcontractors to discuss common problems and to obtain cooperative agreements (see Chapter 2).

Even with efficient and successful expediting, unexpected difficulties can interfere with the schedule progress. Group conferences with subcontractors can help accomplish the following:

- Reveal the worries and concerns of subcontractors.
- Encourage their constructive criticism.
- Elicit suggestions for saving time and money.
- Improve and expand the sharing of materials and equipment.
- Create an understanding and encourage subcontractors to voluntarily start sooner, speed up, slow down, or delay as necessary.

Subcontractors who have not begun work, but who are scheduled to start soon, may be invited to attend meetings. A record is kept of the persons who attend, and minutes are recorded. In very complicated projects, these meetings can be vital to the smooth flow of progress. The superintendent might mail a copy of the minutes to every subcontractor connected with the project, whether or not every subcontractor has attended the meetings.

5. The superintendent can now break down the estimated labor cost into man-hours or crew-hours in order to budget the employee time. As an example, the budget might contain:

	Quantity	Unit Price	Labor
Fine grade for slabs	20,000 sf	$ .11	$2,200
Form footing sides	1,600 sf	$1.35	$2,160
Place concrete	45 cy	$7.23	$ 325

If the pay scales are:

Carpenter—$20.55/Hour, and Laborer—$16.10/Hour, then:

Fine grading

$$\frac{\$2,200}{\$16.10/\text{man-hour}} = 136.6 \text{ man-hrs, or a 4-man crew for 34 hrs.}$$

Forming

$$\frac{\$2,160}{\$20.55/\text{man-hour}} = 105.1 \text{ man-hrs, or a 6-man crew for 17.5 hrs.}$$

Placing Conc.

$$\frac{\$325}{\$16.10/\text{man-hour}} = 20.2 \text{ man-hrs, or a 5-man crew for 4 hrs.}$$

Since the budget was only an estimate, these calculations can be rough; but they are useful in making up a manning chart (see Chapter 15, "Productivity and the Manning Chart").

Given the budget in our example, it is clear that four workers must complete 20,000 square feet of fine grading in three days, or 1,667 square feet per worker per day.

The same restriction on time applies to the forming and placing of concrete. An item of work which will take many days to complete may be cost controlled by dividing it into quantities for daily production.

Since layout and prefabrication comprise a large part of formwork, judging production on the basis of in-place formwork is difficult; costs may appear deceptively high. A close approximation of real cost may be obtained by the use of a formula. For example, let:

Prefabricating   = 40% of the total forming cost
Erecting in place   = 40% of the total forming cost
Stripping & cleaning = 20% of the total forming cost

In our example, on the 1st day a 6-man crew should have completed: 1,600 sf divided by 105.1 man-hours, times 6 men, times 8 hours/day, times 1 day—which equals 730 square feet of forming (which is within the 17.5 crew hours cost budget). But suppose that we find the following actual performance:

Prefab & in-place	500 sf @ (.40 + .40)	=	400 sf
Prefabricated only	750 sf @ .40	=	300 sf
Stripped & cleaned	0 @ .20	=	0
Equivalent of total in place			700 sf

This is below the desired production rate, and the superintendent will need to speed up the crew performance.

Chapter 15

Productivity and the Manning Chart

The term "productivity" as used in this book refers to the ability of men and equipment to produce a certain quantity of work (output) in a certain unit of time (hours, days, weeks). The term "production" refers to actual quantity produced in a unit of time. Unless stated otherwise, production means "rate of production", and a production such as 100 square feet per hour (100 sf/Hr.) means 100 square feet per worker per hour, or 100 square feet per machine per hour.

A cost estimate is based on assumed productions, examples of which have been computed and printed on the cost record system sheet in Chapter 13, Figure 13.4. These assumed productions are goals for the superintendent. The graph of each item is unique. Figure 15.1 is a type of graph which could apply to any work measured in square feet and having an hourly pay rate of $20/Hr. Notice that as production increases, the unit cost curve tends to level off. Beyond 200 sf/Hr., further increases in production produce very small decreases in unit prices. Economists would call this *the law of diminishing returns*. If the estimator had budgeted $.20/sf, a vertical line projects downward to the reading of 100 sf/Hr., which is the necessary minimum production goal. Other points on the curve would require other rates of production; thus, $.15/sf would require 125 sf/Hr.

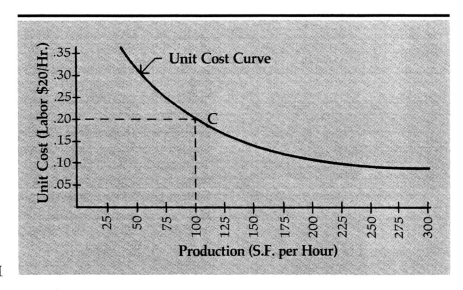

Figure 15.1

When we speak of unit prices and production we mean averages. At the start of a work item, production is low because the worker is unfamiliar with it. Over time, repetition and more efficient management contribute to increased production. These stimuli (repetition and better management) are made possible by larger quantities; small quantities cannot generate enough work time for these influences to have the effect of reducing unit costs. This concept is illustrated in Figure 15.2. Generally, the initial portion of a work item is done at higher unit costs. In Figure 15.3 as production rises to a higher level at P2, unit costs become lower. While P2 is the ultimate goal, P-AVG is the expected average production rate when the early period of productivity is taken into account. The average unit cost, point UP-AVG, combines the early, lower output period (at Start, P1) and the later, higher output period (at P2). While the unit cost curve is negative (down sloping), the production curve is positive (rising to the right). The production curve is also subject to the law of diminishing returns. Figure 15.4 shows both curves on the same graph. Ideally, where they intersect, point UP-AVG represents the point beyond which it is questionable if the rewards in terms of lowered unit costs are worth the effort required to increase production.

What if the estimated unit price (budget) does not coincide with point UP-avg? In Figure 15.5, the unit cost was budgeted at U1 by the estimator, calling for a production of P1; but the superintendent was unable to accomplish a greater production than point P2. The result would be a higher unit cost than estimated for U2, and a cost overrun.

In Figure 15.6, the unit cost was budgeted at U1 by the estimator, calling for a production of P1; but the superintendent was able to accomplish the greater production at point P2, with a resulting unit cost of U2 and a cost savings.

In the case of Figure 15.5, the budget unit cost U1 was too low to be met by normal field production P1. What steps might the superintendent take to push production past point UP-AVG, into the area of diminishing returns?

In Chapter 4, it was pointed out that "pushing" workers is virtually an obsolete practice. Workers are regarded as craftsmen. Estimators tend to budget labor (and equipment) costs at the average production levels which will probably be reached, based on historical performances as found in cost record data. Workers perform in crews at a status quo pace, pulling upward the lower performers and holding back the higher. How then, can productivity be increased? The following paragraphs explore some possible methods.

1. Separate the higher production workers from the "status quo" crews and let them work as teams or independently. Note that high-production workers do not significantly raise status quo production; therefore, their separation is not a loss to the crew.

2. Provide incentives; this is a "pulling" rather than a "pushing" technique. Some incentives are: a raise in pay, bonus, profit from piece work, gifts, and fringe benefits, in addition to the standard benefits, such as a company vehicle for travel to and from the job site and assurance of retaining their jobs during slow downs.

3. Strive for optimum crew sizes, rather than an excess number of workers, to avoid wastefulness and inefficiency. Toward this end, the *manning chart* is helpful. A manning chart is a way of maintaining a constant relationship between the daily hiring of workers and the estimated labor cost budget. It promptly answers the question: "How are we doing, regarding the total labor expenditures?"

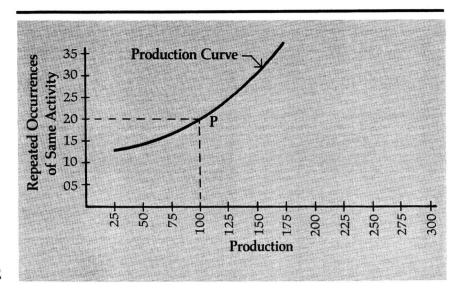

Figure 15.2

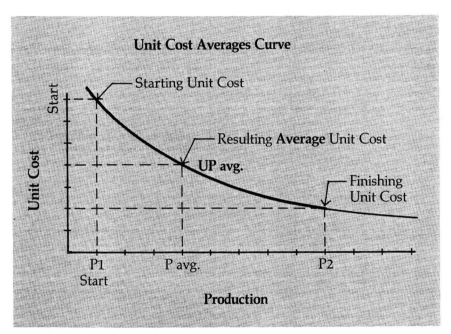

Figure 15.3

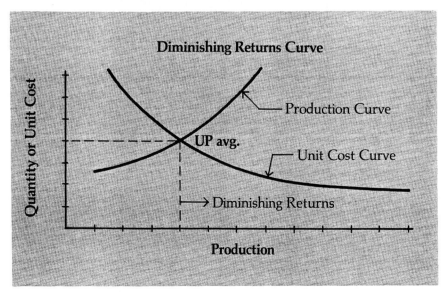

Figure 15.4

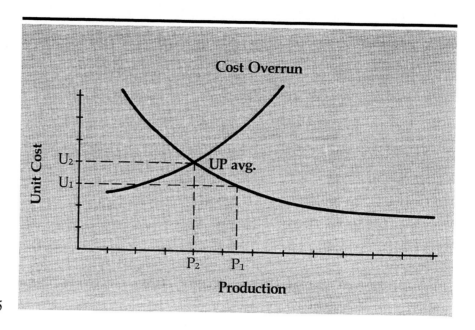

Figure 15.5

4. Most importantly, plan carefully and follow through on details to provide the materials and tools needed to start and maintain crews in motion. This approach minimizes lost time, and thus improves production.

Figure 15.7 is a manning chart constructed from the data previously grouped in the progress schedule (Figure 14.1) and the cost record system (Figure 13.4). The total labor budget in round figures is $107,000. The work force is shown as weekly averages of 2 to 6 workers. The total estimated expenditure in dollars is plotted by a dashed line. Using this chart, the superintendent may quickly and easily determine the status of the actual expenditure, compared to the budgeted figure at any chosen date. For instance, if on May 1st the accounting department reports a total labor expenditure to date of $52,672, the superintendent has the satisfaction of knowing that the cost expenditure (or cash flow) is running under budget by approximately $3,000.

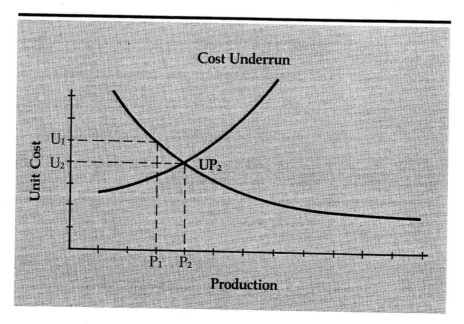

Figure 15.6

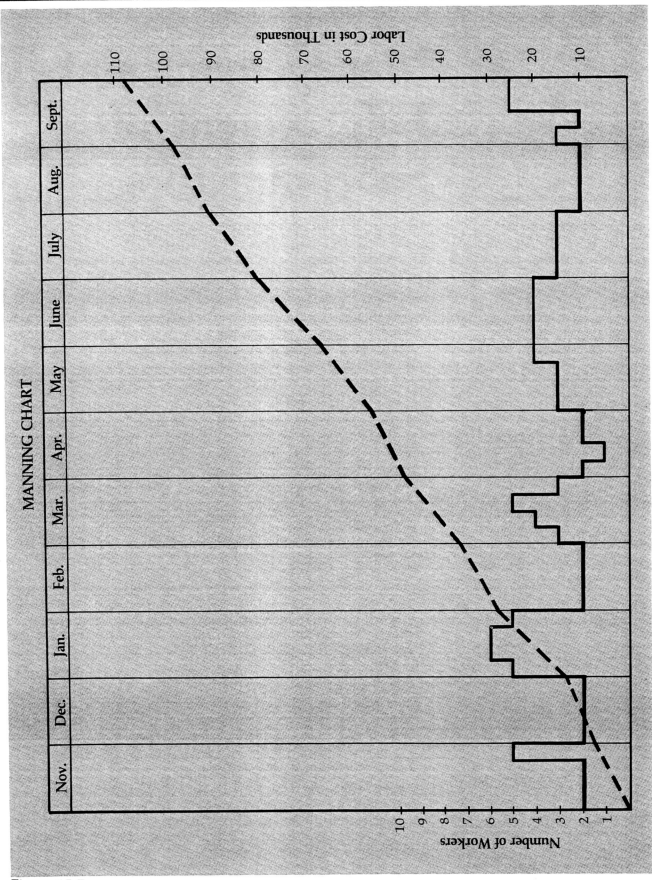

Figure 15.7

Chapter 16

Management and Enforcement of Subcontractors

Problems that require contract enforcement cannot be eliminated, but they can be minimized by good management methods. Several of these methods were described earlier in this book, including:

Chapter 2—explains the superintendent's **management of the forces already in motion** to prevent problems and achieve maximum speed and efficiency. Directing the movement of men and equipment is only one aspect of the management of subcontractors.

Also in Chapter 2 is a section describing **the superintendent's role as inspector**. This aspect of the job requires the superintendent to continually check the materials and workmanship of subcontractors prior to the inspections made by the sponsor and by government agencies. The superintendent's inspections, unlike those of official inspectors, are a form of management.

Administrative Meetings, formal scheduled meetings with subcontractors, are covered in Chapter 11. These meetings contribute to a very effective management approach to complex projects where the completion time period is tight. Subcontractors whose work is currently critical to the project are requested in writing to attend; those whose work is of fringe importance are also invited in writing; all other subcontractors are welcome to attend. The minutes of the meeting are documented; names of those attending are recorded, and a copy of the document is later mailed to the office of each attending subcontractor.

Apart from offering technical enlightenment, these meetings also improve general communications, answer questions, solve problems, and produce agreements. Figure 16.1 is an example of documentation from such a meeting. It may seem trivial standing on its own, but when combined with a chain of such documents, presents evidence of cooperation and teamwork, or the lack of it. This kind of evidence can influence future dealings with particular subcontractors.

Daily reports (Chapter 10), like meetings with subcontractors, serve as another helpful management tool. They not only record current accomplishments, but forecast and help to head off future problems. Company executives who review daily reports are able to contribute their administrative perspective and assistance.

MEETING NOTES

PROJECT Admin. Bldg. & F.S. MEETING # 4 DATE Dec. 20, 1986

PRESENT	COMPANY REPRESENTED
B. Bender	Betbar
P. Piper	Piper Piping Co.
E. C. Static	Sparks Electric Co.
B. Curtis	Histack Masonry

NOT PRESENT

E. Stewart   Utility Plumbing Co.

MAIN BUSINESS TOPICS

1. Speeding up the underfloor utilities

2. Speeding up the submittals of reinforcing steel shop drawings.

RESULTS

Sparks Electric agrees to install sleeves for floor outlets so that the concrete floor slab can be poured; then place outlets and grout around them after approval of submittals are received.

Piper Piping will complete their work two days from this date.

Bender promises to hand-carry rebar shop drawings to the A&E office this day.

OTHER BUSINESS

Histack Masonry agrees to place sleeves, if furnished by others. Piper Piping agrees to furnish the sleeves.

By _Buck Buttress_

Figure 16.1

Watching the weather (Chapter 10) and making decisions about limiting or calling off work is another form of subcontract management. Subcontractors tend to leave such decisions to the project superintendent. When rain is predicted, the superintendent might instruct outdoor workers, such as pipe layers, not to report for work, while permitting indoor workers, such as electricians and finish carpenters, to report as usual. Nevertheless, subcontractors are independent contractors, and they may, at their own risk, attempt to perform their work.

Sociological considerations (Chapter 4) also play a part in the management of subcontractors. Very few relationships occur with a mere oral agreement and signatory handshake. The complexities of modern life preclude such happy arrangements. In general, contractors regard subcontractors as transitory. Thus, precisely worded subcontracts are essential.

Some construction professionals believe that subcontractors should be dealt with impersonally and equally. Others think that subcontractors who repeatedly present financial, technical, or management problems do not deserve equal consideration, and repeat business with them is not desired. Relatively trouble free subcontractors tend to establish preferred status with the superintendent and his company.

Sponsor relations (Chapter 5), whether hard-nosed or congenial, also affect the performance of subcontractors and the superintendent's managment methods.

Selecting methods of construction is one of the early operations in the project to concern the superintendent. Sometimes it is a joint enterprise between the superintendent and subcontractors. Following are some examples of specially chosen construction methods:

- A common trench might be used for utility lines (steam, air, electrical). This shared usage saves time, cost, and street repairs. Subcontractors share the cost and the savings.
- Scaffolding and hoisting equipment might be shared by subcontractors and the general contractor.
- Concrete block work might be delayed at various elevations for the pouring of concrete slabs and other access needs.

In the original bidding and planning, subcontractors may anticipate the methods of construction most likely to be chosen by the superintendent, while remaining open to innovation. Subcontractors are a good source of advice to the superintendent who is wise enough to tap that reservoir of experience.

When **planning the working area** (Chapter 6), the superintendent may reserve space for subcontractors to store materials and equipment and in some cases, office trailers. The superintendent shows good management by making it possible for subcontractors to organize their activities for best efficiency. In some projects one or more subcontractors have a majority of the project work (steel, masonry, mechanical, or electrical), and deserve a proportionate majority of the available work area.

General conditions (Chapter 20) are a common responsibility of the contractor and subcontractors. It is customary for subcontractors to contractually share general conditions costs and responsibilities, usually defined in the project specifications. However, when subcontractors are not specifically named in the subcontracts, enforcement may be difficult and subcontractor relations strained. Some typical general conditions shared

with subcontractors are:
- layout work
- utility charges
- rental equipment
- transportation
- cleanup
- temporary water, heat, power
- testing and inspections
- dewatering

Construction progress schedules are important management tools that continually affect the performances of subcontractors. These schedules, based on legally enforceable provisions, allow little leeway or independence for subcontractors. Instead, subcontractors are managed by the superintendent as team members.

During the bidding stage, subcontractors are carefully selected based on current economic conditions as well as their reputations as team players. During the construction stage, subcontractors are rated by the superintendent in terms of their desirability for repeat business. Poor performers are not necessarily rejected from future subcontracts, but when these workers are hired again, contingency allowances might be included in the bid estimate to cover supporting costs.

The superintendent, more than any other person, establishes the level of quality in a project. There are, of course, many other contributors, but the superintendent has the authority to raise, lower, or accept the standards of quality as they evolve daily on the job. The final credit for job quality tends to go to the management skill of the superintendent.

Change orders (Chapter 17) comprise one aspect of the management of subcontractors. The goal in these dealings is to blend the changed work into the original work as smoothly as possible, with minimum delay. Success depends on prompt transmittals of drawings, specs, prices, approvals, and authorizations. The superintendent must urge both the subcontractors and the sponsor to expedite change orders.

Progress payments (Chapter 18) provide an important management opportunity. Not only does the contractor have the power to "control the purse strings", but prompt and adequate payments to subcontractors is incentive for excellent performance and repeat business. Subcontractors are inclined to accept a more hard-nosed approach if monthly payments are dependable.

On the other hand, when subcontractors (1) repeatedly fail to keep their promises, (2) fail to pay their material suppliers and/or workers, or (3) fail to keep up with the progress schedule, the superintendent may recommend (to his own company manager) withholding part of the subcontractor's payment. This amount reflects the estimated cost of a replacement subcontractor. The general contractor's right to protect his interests from a defaulting subcontractor is often spelled out in a well written subcontract.

Enforcing Subcontractors

It is not uncommon for a subcontractor to present the superintendent with a problem more difficult and demanding than is typical of his daily directing duties. Most subcontractors' workers know what to do and how to do it, and need only the superintendent's directions of when, where, and how fast. There may be some exceptions, however. For example, incompetent workers sent to the project site by a subcontractor can be disruptive to the construction progress schedule and costly to the contractor. This situation calls for remedial action. Other special problems requiring subcontract enforcement are:

- Failure to submit samples, shop drawings, etc.
- A work force and/or equipment that is inadequate in numbers and capacity.
- Repeated breaking of construction time and manpower commitments.
- Refusal to perform certain items which the superintendent believes are the subcontractor's contract responsibility, such as cleanup work, hoisting, scaffolding, etc.
- Ignoring, or making no attempt to comply with progress schedule requirements.

Solving the problem of incompetent or too few workers logically begins with the voicing of a complaint. A strong enough demand might result in the appearance of more competent workers on the job site. A strong economy can put a further strain on the supply of competent workers. In such cases, it may become necessary to make demands and to be persistent. A common failure of this approach occurs when a subcontractor begins a chain of broken promises, rather than complying promptly with the superintendent's requirements. This response is usually a signal that the subcontractor is, in fact, unable to perform.

When a subcontractor refuses or fails to perform work which the superintendent believes is the subcontractor's responsibility, there may not always be time for extended debate. In this situation, the superintendent might choose to have the work done by others and backcharge the subcontractor by deducting the cost from his balance due.

Prompt submittals for approval, the furnishing of competent workers in sufficient numbers, the performance of specific contract items, the negotiating of change orders, and cooperation and compliance with the construction progress schedule—these are all contractual obligations, and failure in one or more of them may be grounds for cancellation of the contract. Cancellation is a last resort, taken only when all enforcement techniques fail. A list of these techniques, from the mildest to the most drastic, might look like this:

1. Voicing a complaint.
2. Withholding a moderate portion of the payment.
3. Paying others to do the disputed or refused work and backcharging the exact cost to the subcontractor.
4. Withholding an excessive and punitive amount of the payment.
5. Cancelling the subcontract and withholding a sufficient amount from the subcontractor's final payment to cover extra costs.

It is possible for each of these techniques to be used to excess. Verbal complaints, for example, can turn into abuse; withholding payments can become a form of revenge; backcharging can be out of proportion; cancelling of subcontracts can be precipitous. On the other hand, too much hesitation by the superintendent can compound the problem. Cancelling subcontracts is almost a sure loss of time and money. Consequently, superintendents tend to procrastinate, prolonging the use of other, less drastic and often ineffective methods.

Chapter 17

Letting Subcontracts and Processing Change Orders

In Chapter 6, four work orders were shown as examples. These are the kind of simplified subcontracts that superintendents issue when it is advantageous to employ independent contractors. When the superintendent first receives the project package (see Chapter 9—"Introduction to the Project"), he finds that a majority of the trades have already been sublet by formal contract. The remaining trades are his direct, personal responsibility. In some cases, he decides rather than purchase materials and hire workmen, to contract with specialists who possess the proper tools and skills. Examples of the kinds of specialty items or trades are:

- Equipment work, such as backhoe, crane, dozer and trucking
- Surveying
- Cleanup: trash pick-up, janitorial service, window cleaning
- Concrete saw cutting and core drilling
- Soil testing
- Dewatering
- Drilling: for piers, jacking under pavement
- Special concrete formwork
- Concrete hoisting or pumping
- Concrete finishing and curing
- Entire concrete work: walks, curbs, precast items
- Caulking
- Hanging doors, installing hardware
- Installing building accessories

Although work orders are more informal than regular subcontracts, the superintendent should be familiar with, and methodically follow certain guidelines in order to avoid legal or economic problems. The following list is a suggested guideline for information to be included in subcontracts:

- Name of project
- Name of independent contractor
- Date of work order
- Nature of agreement, such as to furnish labor only, material only, equipment only, or any combination of these
- Description of work and/or materials

- Specific location of work on site
- Statement of commitment to the original drawings and specs
- Special inclusions and exclusions
- Completion date
- Dollar amount
- Special terms, such as taxes, insurance, basis of payments, etc.
- Signatures and titles under company names

Some contractors set a limit (such as $5,000) for any single work order. Work exceeding that amount is routed through the contractor's main office for formal subcontracting procedure.

Change Orders

The superintendent should have the ability to do cost estimating work. While complex change orders are usually turned over to the company estimating department along with pertinent facts and information, the superintendent should be able to analyze the figures produced by that department. It is important to the superintendent that change orders are quickly calculated and submitted for sponsor approval as this will avoid slowdowns in the progress of the project.

Sometimes change orders require the demolishing and rebuilding of portions of the project and involve a wait for materials not readily available. The contractor and subcontractors are hesitant to order materials and do extra work before firm, written agreements are made with the sponsor; and the sponsor is reluctant to okay extra work before the price is known. A certain amount of urgency is, therefore, normal in change order processing. Because of the time it takes to process, as well as the extra time for accomplishing the change order work, extensions to the original construction time may be justified, and the progress schedule altered. Requests for time extensions, if justified beyond the original contract commitment, should always be included in the change order request.

Time is a crucial factor in a change order, and the most complex aspect of change order estimating usually involves the subcontractor. Therefore, the first order of business is to determine which subcontractors are affected. Those who are must be supplied with drawings, specs and/or detailed descriptions of the change order work. They should also be given deadlines for the submittal of written quotations. Figure 17.1 is an example of a letter of transmittal to a subcontractor requesting a price for change order work.

After price requests have been made to all affected subcontractors, the superintendent (or company estimating department) must estimate the cost of the remaining work to be done by company employees. (A complete sample change order appears in Figure 17.6.) Since all construction trades and a variety of combinations are involved in different change orders, we will briefly review the most common situations found in the general contracting business.

Considering competition: Unlike bidding for the original job, there is no competition for change order work. There can, however, be strong resistance from the sponsor to the proposed prices. The superintendent must be prepared to prove the validity of those prices.

Divisions of the estimate: A total project is carried out by trades under the heading of the general contractor, and trades accomplished by subcontractors. The distinction is clear in the case of change orders, since original subcontracts have already been let and general contractor work has already been determined. Some change orders are entirely subwork, requiring mere compilation from the superintendent. Most change orders are a combination of both.

TO: DATE:

FROM:

REGARDING PROJECT:

     Enclosed are _____ pages of drawings and specs
describing:

ACTION REQUESTED: Submit in duplicate a firm price with
                       breakdown showing quantities, materials,
                       equipment, overhead and profit for your
                       portion of this extra work, which includes:

SPECIAL NOTES: Furnish quotation no later than _____

                                                   Your Own Construction Company

                                           B. Buttress
                                         Superintendent

Figure 17.1

Sources of cost information: to guide in making correct cost estimates:

1. Records of actual unit costs as repeatedly proved in the field and kept in company files.
2. Tables and mathematical relationships as found in construction text books.
3. Cost reference books, such as R. S. Means' *Building Construction Cost Data*.
4. Constructive imagination based upon the superintendent's experience.
5. Hearsay information from experts.

An immediate source of information is the original estimate for the entire project, and the cost record system (see Chapter 13, Figure 13.4). As a rule, however, change order work warrants higher unit prices than the original for reasons which will be noted below under the heading, "Ethical Considerations".

Forms, formats and systems of estimating: Sponsors often have formats and systems which they prefer for uniformity throughout their projects. The alternative is to use the forms offered by individual contractors. When the sponsor does not otherwise specify, the superintendent may follow his own company's system.

Interpretations and disputes: Occasionally a sponsor will demand work which the superintendent believes is not in the contract (N.I.C.). The work might fall into a gray area, not clearly indicated in the drawings and specs; or, it may not have a precedent in custom-of-the-trades. The sponsor might insist that the work is the contractor's responsibility. A good way for the superintendent to deal with this problem is to promptly compile a change order estimate and consult company management. He should not proceed with the questionable work without written instructions from either the sponsor or his own company manager. Another approach is to request from the sponsor a directive in writing: then follow up with a letter agreeing to proceed with the disputed work under protest and with the intention of submitting the facts to the contractually predetermined third party for a ruling (see Chapter 5—"Sponsor Relations").

Ethical Considerations: Unlike the original bidding situation, the lack of competition in change orders can make pricing ethically suspect. A fair price for change order work is somewhat higher than the original work in order to compensate for the overhead costs involved in planning, estimating, dealing with subcontractors, supervising, and processing the paperwork. The markups proposed by the sponsor may not be adequate compensation to the contractor. The example change order at the end of this section shows a common method for arriving at an adequate price increase.

Mistakes: The four types of mistakes that affect change orders are (1) judgment, (2) omissions, (3) arithmetic, and (4) slips. Mistakes in change orders due to omissions, arithmetic, or slips are usually found and corrected, since both the superintendent and the sponsor have the opportunity and desire to check the figures and the content carefully. Mistakes of judgment, however, are not easily corrected after the price is formally agreed to by both parties. For this reason, careful consideration should be given to construction methods and the productivity of workers and equipment.

Selecting methods of construction: Since change orders typically consist of the same kind of work already under way, the methods and equipment already being used receive first consideration. In other instances, the original work might already be complete and the additional work required by the change order might involve an entirely different approach (hand work, for instance, instead of machine work). In this case, the cost could be comparatively higher.

Mobilization and time lag allowance in labor: If change order work is well timed, it can be smoothly integrated with the work in progress and thus require no additional mobilization and time lag. However, a delay in processing or sponsor approval might put off the additional work until it is too late to mesh with the original. Since delays are a regular occurrence in change order negotiations, the estimates should contain appropriate cost allowances. Other conditions which might affect costs are those relating to climate or season. If change order work will be done in extremes of temperature, or in rain, it could cost more than the original work.

Keeping up with changes (escalation): In the original estimating and bidding of a project, allowances are made for material, equipment and labor cost increases that might occur in the course of the project. Whatever the method used in bidding, later change orders should incorporate the price levels that are current at the time the change order work will be done. Figure 17.2 shows the selection of labor pay scales. Material and equipment prices are a matter of obtaining current quotations from suppliers. Shown below is an eighteen month project subject to two scheduled wage increases. A total of three wage levels are involved. Change order (C.O.) A occurs during the first level and is priced accordingly, C.O. B overlaps into the second, higher wage level; C.O. C falls entirely within level two; C.O. D overlaps into level three; and C.O. E is entirely within level 3. Clearly, each change order would be priced slightly higher than the one before.

Change orders should be neatly typed, preferably on pre-printed forms, and should be supported with reference data and explanations. Change orders will be critically reviewed and studied in detail by the sponsor.

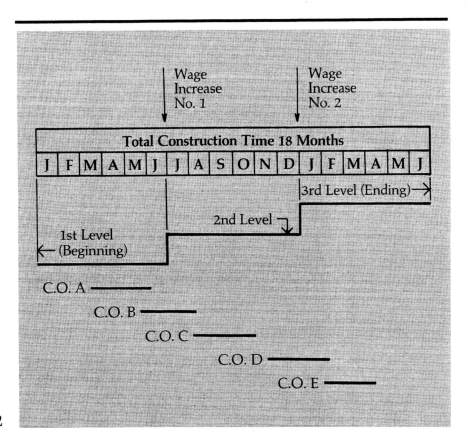

Figure 17.2

Taking off quantities: Quantities are easily checked by the sponsor and therefore should be correct and beyond dispute. Some quantities may be transferred directly from the drawings to the price-out sheet, while more complex change orders might warrant the use of quantity survey sheets (such as in Figure 17.3).

An advantageous quantity takeoff method is one in which all elements associated with an item are kept together in a group. For instance, lined up under **Footings** might be: layout, machine excavation, hand excavation, formwork, concrete, backfill, and disposal of excess dirt. Lined up under **Concrete Slab on Grade** might be: fine grading, gravel base course, waterproof membrane, edge forms, setting of screens, concrete, finishing, and curing.

General Conditions: Change orders that do not require time extensions beyond the established completion date need fewer general conditions cost items. Even these kinds of change orders might involve some extra expenses, however, such as surveying, layout, cleanup, and scaffolding. Change orders that do require extensions beyond the completion date may include many of the typical monthly costs, such as office, telephone, utilities, toilets, transportation, and supervision (see Chapter 20—"General Conditions").

When a prospective change order approaches a critical date, the superintendent might include a contingency clause in the quotation. For example:

"We propose to do the above described extra work for the price of $_____ contingent upon receipt of your authorization to proceed before _____ (date). After _____ (date) our price would be increased by $_____ to cover general conditions costs for two weeks of extension to the original construction completion time."

Demolition: Since all demolition work is usually completed before change orders occur, it may be uneconomical to call the subcontractor back to the site for minor extra work. The superintendent, therefore, estimates the cost and plans to accomplish the demolition with company workers. The superintendent may rely on personal knowledge and experience to determine expected production of workers and equipment. Published cost reference data serves as a check to confirm the superintendent's estimated unit prices. This is a way to quickly, but roughly, focus on the target; a more accurate adjustment can then be made by using firm material quotations, actual labor pay scales, true equipment rental rates and a realistic allowance of hours to do the work.

Fringe benefits and payroll taxes: As a rule, estimates are calculated from the "base pay", or the gross amount of paychecks given to the workers. When all of the estimated labor costs are summarized into one total figure, an additional amount must be computed to cover the various taxes, insurance, and pensions which the company must pay to agencies on behalf of the employees. The superintendent does not need a detailed breakdown of these expenses; the home office can supply him with a current value in the form of a percentage of the total labor cost. For our example in this book, let us use 47.6%.

Overhead and Profit. In the following example, let us assume that 10% plus 8% is the maximum markup permitted in the contract documents written by the sponsor for his protection. From the contractor's viewpoint, this markup is inadequate for change order work. An alternative is to increase the unit

PROJECT Administration Bldg. & Fire Station

ESTIMATE NO.

LOCATION Picacho, CA   ARCHITECT   DATE October 10, 1987

TAKE OFF BY BB   EXTENSIONS BY: BB   CHECKED BY: PC

DESCRIPTION	NO.	L	W	D		UNIT		UNIT		UNIT		UNIT
Demolition												
1a Sawcut Wall	34	—	—		—	34	LF		—		—	
1b Breakout Wall	—	3	7	21	SF		—		—		—	
1c Sawcut Floor	24	—	—		—	24	LF		—		—	
Concrete												
2a Excavation	11	4	0⁶		—		—		—		—	
2b Formwork	24	—	—		—	24	LF		—		—	
2c Concrete	11	4	1²		—		—	53	CF		—	
2e Finish/Cure	11	4	—	44	SF		—		—		—	
					44	SF	24	LF	53	CF (2 CY)		—
2d Expan. Jt/Sealer						—	24	LF		—		—
2f Anchor Bolts						—		—	6	Ea		—
Rough Carpentry												
3a Wall Framing		448	—	2×4		—	300	bf		—		—
3b Ceiling Framing		164	—	2×8		—	220	bf		—		—
3c ½" Plywood	5	4	×	8	160	SF		—		—		—
3d Rough Hardware		680 bf × .01				—		—	7	$		—
Finish Carpentry												
4a Door Frame		—	—	3⁰ 7⁰		—		—	1	Ea.		—
4b Door	SC 1¾	—	—	3⁰ 7⁰		—		—	1	Ea.		—

Figure 17.3

prices, or to further break down the change order, specifically spelling out some of the overhead items as individual work items and then applying the standard markups to that new total.

Bond Cost: Construction bonds are a form of insurance protecting the sponsor from contractor's defaults. Since the cost of bonding is not standard between contractors and their bonding companies, the superintendent needs to know his own company's formula. Ordinarily, the rate decreases as the size (cost) of a project increases.

Figure 17.4 is an example of a change order request from the sponsor; Figure 17.5 is the detailed cost estimate; Figure 17.6 is the letter of proposal (offer) sent to the sponsor in response to his request.

Complicating this change order is the element of time. Since no extension is allowed, and the project is very nearly completed, overtime work is unavoidable. the affected subcontractors are notified of the time restriction and are allowed to judge for themselves if overtime is necessary for their respective portions. The superintendent itemizes each element of the work, beginning with demolition. Quantities are listed, but because of the small volume, the average unit prices are inaccurately low. Consequently, man-hours are judged and unit prices computed by the following formula:

$$\text{Unit Price} = \frac{\text{man-hours}}{\text{per unit}} \times \frac{\text{pay rate}}{\text{per hour}}$$

The general contractor's work is estimated first. In this example it consists of demolition, concrete, and rough and finish carpentry. The cost of overtime work is shown as a separate item (# 5) in order to call the sponsor's attention to the consequence of his time restriction.

Subcontractor trades are then entered in the exact lump sum amounts of their quotations. Each of their detailed breakdown estimates is submitted to the sponsor along with the superintendent's paperwork.

Fringe benefits are then added on. They amount to 47.6 of the general contractor's total estimated cost for basic labor.

The permissible overhead and profit is computed and added on. The cost of the bond is also calculated and included to complete the change order estimate.

Note: Different markup percentages may sometimes be specified for the work subcontracted and the work to be performed by the general contractor's own work force.

                              Date

Your Own Construction Co.
Address
City, State   zip code

ATT: Company President

RE: Administration Building & Fire Station
      Picacho, California

Dear Mr. Keefe:

    Please submit a complete breakdown of quantities, unit
prices and extensions for all labor, material, equipment and
subcontract work to construct a generator room in accordance
with the specifications below and the pertinent sections of
the contract specifications.

    Construct two walls 10'-0" long and one wall 16'-0" long.
Wall to be 2 x 4 wood studs, 16" o.c. Wall to be 10'-0" high.
Exterior of wall to receive 5/8" gypsum board. Interior of walls
and ceiling to be metal lath and plaster. Cut into existing
concrete block wall a new 3'-0" x 7'-0" x 1-3/4" Hollow Metal
"B" Label door and P.M. frame.

                              Very truly yours,

                              Bell & Bell, Architectural Assoc.

                              Chris Bell, Manager

Figure 17.4

Means Forms
COST ANALYSIS

PROJECT: Administration Bldg. & Fire Station   CLASSIFICATION

LOCATION: Picacho, CA   ARCHITECT

TAKE OFF BY: PC   QUANTITIES BY: PC   PRICES BY: AT   EXTENSIONS BY: AT   CHECKED BY: JM

ESTIMATE NO.

DATE: 10-13-87

Description	Quantity	Unit	Material Unit Cost	Material Total	Labor Unit Cost	Labor Total	Equipment Unit Cost	Equipment Total	Subcontract Unit Cost	Subcontract Total	Total Unit Cost	Total
1. Demolition												
a. Saw Cut Wall	34	L.F.	.32	11	2.40	82						93
b. Breakout & Remove	21	S.F.	0		.99	21						21
c. Saw Cut Floor	24	L.F.	1.28	31	2.84	68						99
d. Breakout & Remove	32	S.F.	0		1.72	55						55
2. Concrete												
a. Excavate Hand book	1	C.Y.	0	0	16.10	16						16
b. Formwork	24	S.F.	.27	6	1.35	32						38
c. Concrete, Furnish, place	2	C.Y.	50.90	102	7.23	14						116
d. Expansion Jt & sealer	24	L.F.	.75	18	.82	20						38
e. Finish & cure	44	L.F.	.05	2	.33	15						17
f. Anchor bolts	6	Ea.	.55	3	1.93	12						15
3. Rough Carpentry												
a. Wall Framing	300	BF	.34	102	.59	177						279
b. Ceiling Framing	220	BF	.34	75	.28	62						137
c. Plywood ½"	160	S.F.	.39	62	.25	40						102
d. Rough hardware	1	L.S.	7	7								7
4. Finish Carpentry												
a. 3/0 7/0 Door Frame	1	Ea.	.29	29	.16	16						45
b. S.C. Door 3/0 7/0	1	Ea.	86	86	.27	27						113
5. Overtime	72	Hrs.			10.25	738						738
Subcontract Quotes												
Reinforcing Steel	Quote									96		96
Finish Hardware	"									70		70
Gypsum Board & Plaster	"									1003		1003
Painting	"									751		751
Mechanical	"									1500		1500
Electrical	"									1000		1000
Subtotals				534		1395				14420		16349

Figure 17.5

Means Forms
COST ANALYSIS

PROJECT: Administration Bldg & Fire Station

LOCATION: Picacho, CA

TAKE OFF BY: PC   QUANTITIES BY: PC   PRICES BY: PC

CLASSIFICATION

ARCHITECT

EXTENSIONS BY: AT

CHECKED BY: JM

ESTIMATE NO.

DATE: 10-13-87

Description	Quantity	Unit	Material		Labor		Equipment		Subcontract		Total	
			Unit Cost	Total	Unit Cost	Total	Unit Cost	Total	Unit Cost	Total	Unit Cost	Total
Carried Forward				534		1395		—		14420		16349
Labor Mark-up 47.6%						664		—				664
												17013
Overhead 10%									Subtotal			1701
												18714
Profit 8%									Subtotal			1497
									Total			20211

Figure 17.5 continued

93

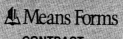 **Means Forms**

**CONTRACT
CHANGE ORDER**

FROM:

_____

_____

_____

TO:

_____

_____

_____

_____

CHANGE ORDER NO. _____

DATE _____

PROJECT _____

LOCATION _____

JOB NO. _____

ORIGINAL CONTRACT AMOUNT $		
TOTAL PREVIOUS CONTRACT CHANGES		
TOTAL BEFORE THIS CHANGE ORDER		
AMOUNT OF THIS CHANGE ORDER		
REVISED CONTRACT TO DATE		

Gentlemen:

This CHANGE ORDER includes all Material, Labor and Equipment necessary to complete the following work and to adjust the total contract as indicated;

☐ the work below to be paid for at actual cost of Labor, Materials and Equipment plus _____ percent (_____%)

☐ the work below to be completed for the sum of _____

_____ dollars ($_____)

_____

_____

_____

_____

_____

_____

_____

_____

_____

_____

_____

_____

_____

_____

_____

_____

_____

_____

_____

_____

_____

CHANGES APPROVED

The work covered by this order shall be performed under the same Terms and Conditions as that included in the original contract unless stated otherwise above.

By_____

By_____

Signed_____

By_____

III.63

Figure 17.6

Chapter 18

Progress Payment Requests and Backcharging

As work proceeds, the contractor and each of the subcontractors spend their own money (operating capital) on payrolls, materials, and equipment. Soon, usually in one month, they need to be reimbursed. Each periodic payment contains a markup for overhead and profit.

Immediately after the general contract is signed, a master progress payment schedule form should be presented to the sponsor for approval. This schedule serves as a basis for computing monthly billings by the contractor. Figure 18.1 is an example of such a form.

Each month, prior to an established deadline date, the superintendent must calculate the amount of work actually performed to date, as well as the dollar value (based on the unit prices in the master schedule). Figures 18.2 and 18.3 are examples of monthly payment requests to compensate for work already performed.

If the master schedule has been well designed, each monthly progress payment request may be a straightforward, routine report of actual material quantities and work performed. There should be little conflict with the sponsor's representative regarding these values; furthermore, overstocks and shortages tend to be self-correcting from month to month.

Before making the formal payment request, the superintendent should ask subcontractors for their own assessments of completed work. This approach may help to avoid later arguments. All other items (not sublet) are calculated by the superintendent. The breakdown is then discussed with the sponsor's representative. The representative's agreement is strong assurance that the request will be accepted by the sponsor and that payment is forthcoming.

An occasional complication may arise when a subcontractor insists on billing for a larger amount than the superintendent or representative believes is justified. The contractor is careful not to overpay the subcontractors.

Figure 18.2 is an example of a first billing. Only five items are involved, but the total amount in dollars is large, due to the volume of earth moved. Figure 18.3 is an example of the fifth billing, which includes 22 line items; 11 of these items are completed. Future payment requests will indicate increasing percentages of completion. Clerical time can be shortened by inputting this data to a properly programmed computer (see Chapter 10 — "Record Keeping").

Administration Building
And Fire Station   MASTER PROGRESS PAYMENT SCHEDULE

Picacho, California   Page 1 of 2

PAY ITEM	COST CODE	WORK DESCRIPTION	QUANTITY	LABOR U/C	LABOR TOTAL	MAT'L/EQUIP U/C	MAT'L/EQUIP TOTAL	TOTAL	NOTES
100	A100	Move on	1 ls	1s	5,000	1s	4,890	9,890	(1)
120	B100	Surveying	sub	1s	8,000	1s	1,000	9,000	(1)
121	B110	Progress schedule	sub	1s	7,000	1s	500	7,500	(1)
122	A130	Earth shoring	1 ls	1s	5,108	1s	5,092	10,200	(2)
123	A131	Earthwork	45,000 cy	.48	21,600	10.00	450,000	471,600	
130	B111	Ch. link fence	540 lf	8.44	4,558	11.00	5,940	10,498	(1)
131	B112	A.C.paving, base course	2,400 cy	1.75	4,200	12.00	28,800	33,000	(1)
	B113	A.C.paving, finish course	108,000 sf	.12	12,960	2.00	216,000	228,960	
132	A150	Parking striping	3,000 lf	.30	900	.10	300	1,200	
133	B114	Landscaping	sub	1s	3,000	1s	3,000	6,000	
134	B115	Irrigation sprinklers	sub	1s	2,222	1s	2,000	4,222	
135	B116	Reinforcing steel	28,500 lb	.25	7,125	.25	7,125	14,250	
136	B117	Masonry	3,500 sf	5.00	17,500	6.18	21,630	39,130	(1)
137	B118	Structural steel	2,326 ton	25.00	58,150	50.00	116,300	174,450	
	B119	Miscellaneous metal	11,333 lb	.75	8,500	.75	8,500	17,000	
	B120	Steel Decking–2nd floor	2,000 sf	.70	1,400	1.23	2,460	3,860	
	B121	Steel Decking–roof	9,000 sf	1.00	9,000	.93	8,370	17,370	
138	A151	Concrete (except site work)	352 cy	117.24	41,268	107.39	37,801	79,069	
139	A180	Carpentry, rough & finish	1s	1s	10,731	1s	6,766	17,497	
	A181	Install miscellaneous metal	1s	1s	2,004	----	----	2,004	
140	B122	Millwork	sub	----	----	1s	23,000	23,000	
141	A182	Wood doors	18 ea	----	----	70.00	1,260	1,260	
142	B123	Wall & ceiling insulation	12,200 sf	.25	3,050	.25	3,050	6,100	
143	B124	Roof insulation	9,000 sf	.60	5,400	.77	6,930	12,330	
144	B125	Clay tile roofing	9,000 sf	1.16	10,440	3.00	27,000	37,440	(1)
145	B126	Sheet metal	1,420 lb	1.00	1,420	1.00	1,420	2,840	
146	A183	Wall louvers	8 ea	40.00	320	80.00	640	960	
147	B127	Calking & sealants	2,280 lf	1.00	2,280	.55	1,254	3,534	
148	B128	Hollow metal doors & frames	24 ea	----	----	87.50	2,100	2,100	
149	B129	Coiling steel doors	4 ea	500.00	2,000	1700.00	6,800	8,800	
150	B130	Aluminum windows	34 ea	150.00	5,100	166.17	5,650	10,750	
151	B131	Glass & glazing	1,680 sf	1.00	1,680	1.50	2,520	4,200	
152	B132	Finish hardware	sub	----	----	1s	3,870	3,870	
153	B133	Lathing & plastering	2,416 sy	8.00	19,328	7.00	16,912	36,240	(1)

Figure 18.1a

96

Administration Building
and Fire Station
Picacho, California

MASTER PROGRESS PAYMENT SCHEDULE

PAY ITEM	COST CODE	WORK DESCRIPTION	QUANTITY	LABOR U/C	LABOR TOTAL	MAT'L/EQUIP U/C	MAT'L/EQUIP TOTAL	TOTAL	NOTES
154	B134	Metal framing & gyp board	2,777 sy	10.00	27,770	8.00	22,216	49,986	(1)
155	B135	Ceramic tile	550 sf	3.50	1,925	3.50	1,925	3,850	
156	B136	Acoustical tile	3,000 sf	.60	1,800	.60	1,800	3,600	
157	B137	Resilient flooring	8,000 sf	.30	2,400	.43	3,440	5,840	
158	B138	Carpeting	100 sy	6.00	600	10.00	1,000	1,600	
159	B139	Painting	100,000 sf	.17	17,000	.18	18,000	35,000	(1)
	B140	Toilet accessories	20 ea	----	----	61.00	1,220	1,220	
	B141	F.E. cabinets	4 ea	----	----	50.00	200	200	
	B142	Kitchen unit	1 ea	280.00	280	500.00	500	780	
160	B143	Toilet partitions	8 ea	66.25	530	100.00	800	1,330	
	B144	Flagpole	1 ea	180.00	180	800.00	800	980	
161	B145	Venetian blinds	25 ea	30.00	750	50.00	1,250	2,000	
	B146	Room numbers	42 ea	2.50	105	7.50	315	420	
162	B147	Sanitary sewer	sub	1s	4,000	1s	6,000	10,000	(1)
	B148	Storm drains	sub	1s	3,000	1s	3,875	6,875	(1)
	B149	Exterior water	sub	1s	4,250	1s	7,000	11,250	(1)
	B150	Exterior gas	sub	1s	2,875	1s	4,000	6,875	(1)
163	B151	Plumbing	sub	1s	11,000	1s	21,375	32,375	(1)
	B152	Hose reels	sub	1s	625	1s	2,000	2,625	(1)
164	B153	Heating	sub	1s	10,938	1s	20,312	31,250	(1)
	B154	Air conditioning	sub	1s	9,875	1s	12,000	21,875	(1)
165	B155	Electrical, underground	sub	1s	8,000	1s	14,500	22,500	(1)
	B156	Electrical, bldg. rough	sub	1s	20,000	1s	20,000	40,000	(1)
	B157	Electrical, bldg. finish	sub	1s	3,000	1s	7,000	10,000	(1)
	B158	Electrical, parking lot	sub	1s	3,000	1s	5,750	8,750	(1)
166	A184	Window & fixture cleaning	1s	1s	2,200	----	----	2,200	(1)
167	A185	Site concrete work	1s	1s	12,655	1s	7,000	19,655	(1)
		TOTALS			430,002		1,213,158	1,643,160	

NOTES:
(1) Increased 25% for distribution of general conditions, markup and bond costs.
(2) Includes the remainder of the distribution not covered in (1).

LEGEND: Cost Code A – Items to be accomplished by Our Own Company (see cost record sheets).
Cost Code B – Items to be accomplished by subcontractors.

Figure 18.1b

Administration Building
and Fire Station   REQUEST FOR PAYMENT, DEC. 5, 1985
Picacho, California

PAY ITEM	COST CODE	WORK DESCRIPTION	QUANTITY	LABOR		MAT'L/EQUIP		TOTAL	NOTES
				U/C	TOTAL	U/C	TOTAL		
100	A100	Move on	100%	1s	5,000	1s	4,890	9,890	
120	B100	Surveying	85%	1s	6,800	1s	850	7,650	(3)
121	B110	Progress schedule	80%	1s	5,600	1s	400	6,000	(4)
122	A130	Earth shoring	90%	1s	4,597	1s	4,582	9,179	(5)
123	A131	Earth moving	22,500 cy	.48	10,800	10.00	225,000	235,800	
TOTAL AMOUNT EARNED					32,797		235,722	268,519	
LESS 10% RETENTION BY SPONSOR								26,852	
AMOUNT PAYABLE								241,667	

NOTES: (3)   Final grade stakes remain to be done
           (4)   Periodic updating remains to be done
           (5)   Removal of shoring remains to be done

Figure 18.2

Administration Building
and Fire Station, Picacho, Ca. REQUEST FOR PAYMENT, APRIL 5, 1986

PAY ITEM	COST CODE	WORK DESCRIPTION	QUANTITY	LABOR U/C	LABOR TOTAL	MAT'L/EQUIP U/C	MAT'L/EQUIP TOTAL	TOTAL	NOTES
100	A100	Move on	100%	1s	5,000	1s	4,890	9,890	(1)
120	B100	Surveying	100%	1s	8,000	1s	1,000	9,000	(1)
121	B110	Progress schedule	90%	1s	6,300	1s	450	6,750	(1)
122	A130	Earth shoring	100%	1s	5,108	1s	5,092	10,200	(1)
123	A131	Earthwork	100%	1s	21,600	1s	450,000	471,600	(2)
134	B116	Reinforcing steel	100%	1s	7,125	1s	7,125	14,250	
135	B117	Masonry	100%	1s	17,500	1s	21,630	39,130	
136	B118	Structural steel	1,200 ton	25.00	30,000	50.00	60,000	90,000	(1)
137	B119	Miscellaneous metal	8,000 lbs	.75	6,000	.75	6,000	12,000	
	B120	Steel decking, 2nd floor	100%	1s	1,400	1s	2,460	3,860	
138	A151	Concrete	330 cy	117.24	38,689	107.39	35,438	74,127	
139	A180	Carpentry	10%	1s	1,073	1s	677	1,750	
	A180	Install misc. metal	20%	1s	401	-----	-----	401	
162	B147	Sanitary sewer	100%	1s	4,000	1s	6,000	10,000	(1)
	B148	Storm drains	100%	1s	3,000	1s	3,875	6,875	(1)
	B149	Exterior water	100%	1s	4,250	1s	7,000	11,250	(1)
	B150	Exterior gas	100%	1s	2,875	1s	4,000	6,875	(1)
163	B151	Plumbing	50%	1s	5,500	1s	10,688	16,188	(1)
164	B153	Heating (rough-in)	40%	1s	4,375	1s	8,125	12,500	(1)
	B154	Air conditioning (rough-in)	20%	1s	1,975	1s	2,400	4,375	(1)
165	B155	Electrical, underground	80%	1s	6,400	1s	11,600	18,000	(1)
	B156	Electrical, bldg. rough	20%	1s	4,000	1s	4,000	8,000	(1)
		TOTAL AMOUNT EARNED			184,571		652,450	837,021	
		LESS 10% RETENTION BY SPONSOR						83,702	
		AMOUNT PAYABLE						753,319	

NOTES: (1) 25% is general conditions, markup and bond cost distribution
(2) $44,514 of this item is general conditions, markup and bond cost distribution

Figure 18.3

Under a subcontract agreement, the general contractor is obligated to pay the subcontractor an established amount for specific materials and work. When the payments are due, the general contractor might deduct a portion (backcharge) to cover certain loose ends. These "loose ends" represent work that the subcontractor has failed to do, thereby falling short of his contractual obligations (as interpreted by the superintendent). When the superintendent has made such a determination, he notifies the subcontractor of the non-compliance, and informs him that the work will be done by other forces, with an impending charge to the subcontractor's account. This impending charge is known as a "backcharge". From the subcontractor's viewpoint, some backcharge items might be unjustified. Some examples of common backcharge items are listed below:

- Patching and repairing
- Paint touching up
- Cleaning up
- Use of equipment
- Labor
- Materials
- Lights, barricades, and traffic control
- Dewatering
- Hoisting
- Delays
- Liquidated damages
- Telephone and utility charges
- Transportation and hauling

Backcharges can be very damaging to business relationships. The superintendent should, therefore, be sure that these items are well within the scope of the subcontracts and/or customs-of-the-trades. Backcharging trivial amounts for items not clearly the subcontractor's responsibility can be poor economics. Furthermore, payments with backcharges deducted should contain adequate explanations, a detailed breakdown of costs, and copies of invoices. See Figure 18.4 for an example.

Date

Daubers Paint Company
1635 Washington St.
San Diego, CA   92104

ATT: Pete Daubers

RE: Administration Building & Fire Station
     Picacho, CA

Dear Pete:

    Enclosed is your final payment for work completed. The

backcharges are explained as follows:

Agreed to by your foreman, John Spears:

        1. 1/2 of rented scaffold (invoice attached) $229.52

        2. Repair broken concrete curb 22 1f @ $9.00 =   198.00

Other work:

        3. Replacing glass in entry door broken by
               your unloading crew (invoice attached)   110.00

                            TOTAL $537.52

                        Yours very truly,

                        Your Own Construction Company

                        B. Buttress
                        Superintendent

Figure 18.4

Chapter 19

Safety Precautions and Quality Controls, Inspections & Tests

As previously stated, one of the superintendent's responsibilities is ensuring safety while a maximum number of units is moved through a restricted site in a minimum time frame. The superintendent should have full knowledge of the current official safety rules and regulations. Good project specs prescribe in detail such safety measures as barricades, lights, flagmen, shoring of trench sides, railings, nets, and scaffolding.

The superintendent's experience is important, as is the habit of being constantly on watch for dangerous conditions. Following are some guidelines which help to minimize risk to both persons and property.

1. The superintendent or an assistant should have at least basic training in first aid, and encourage workers to report and receive treatment for even minor injuries. A log should be kept for administrative purposes, and a first aid kit should be made available. The superintendent should take the initiative in deciding when injured workers should be sent to a medical center.

2. The superintendent should strictly enforce the wearing of hard hats at all times, not only by workers, but also by visitors to the site. The widespread use of hard hats on construction projects is testimony to the high risk of head injuries.

3. Appropriate equipment (such as boots and gloves, goggles, glasses, shields, and masks) should be furnished for the prevention of eye injuries and respiratory problems. In some cases, the construction of a dust-tight partition is advisable to separate workmen from areas in which the air is not safe to breathe.

4. The superintendent should also observe and correct any potentially dangerous methods practiced by workers. Injuries caused by falls can be minimized by such preventive measures as safety belts, railings, and life nets.

5. The danger of electrical shock is best prevented by a careful investigation of old master plans of the site, and by a consultation with utility companies to locate existing and possibly concealed power lines. The superintendent might also enlist the cooperation of the current project's electrical subcontractor to switch off power temporarily.

6. Measures can also be taken to prevent burns from sources such as hot tools, sun-heated materials, open flames, electricity, chemicals, minerals, and sparks from grinders. Gloves, face shields, and tight-fitting collars and sleeves can be helpful.

Other preventable sources of accident or injury include:
- impure drinking water
- unsanitary conditions
- inadequate lighting
- improperly stacked materials
- cave-ins of trench sides, tunnels, etc.
- exposure to irritant plant materials or chemicals
- falling objects
- explosions
- misuse or malfunction of tools

Following is a list of further preventive measures that can be implemented by the superintendent to remove the risks listed above and to insure safety on the job site.

1. Scheduling of regular meetings between the superintendent and all crew leaders to discuss safety measures and to review various safety rules and guidelines.
2. Furnishing pure bottled water for drinking.
3. Providing well-maintained, sanitary toilet facilities.
4. Providing adequate lighting for all working areas, day or night.
5. Providing fire extinguishers, water, blankets, shovels, and other equipment and materials to put out fires.
6. Supervising the handling, stacking, and storing of building materials for levelness, interlocking of members, bracing, and blocking to avoid collapse.
7. Checking stripped forming materials and other lumber for dangerous projecting nails, and seeing to their disposal.
8. Maintaining cleanup of the project; removing obstacles in the path of traffic.
9. Requiring guard railing around the perimeters of the upper floors and roof.
10. Either shoring the sides of trenches and pits or excavating to safe angles of repose.
11. Keeping all electrical and pneumatic tools in good repair.
12. Carefully researching and investigating for existing underground utilities (especially electrical) before excavating.

Quality Controls, Inspections and Tests

In Chapter 2, it was pointed out that the superintendent must anticipate a great number of quality standards. These standards reflect the demand of the authorized individuals who represent the sponsor and certain socio-political agencies. It is sometimes necessary for the superintendent, on his own initiative, to raise lagging quality by urging on the workers and subcontractors. On the other hand, he may occasionally have to suppress quality standards that exceed the project specs, and custom-of-the-trades requirements. Although there might be exceptions, quality and cost have a direct relationship. Quality and productivity have an inverse relationship. These concepts are illustrated in Figures 19.1 and 19.2.

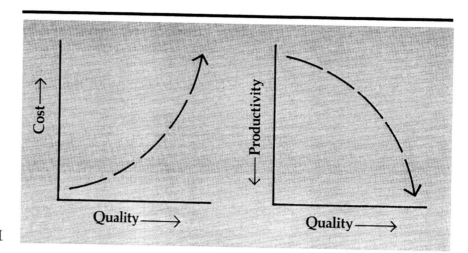

Figure 19.1

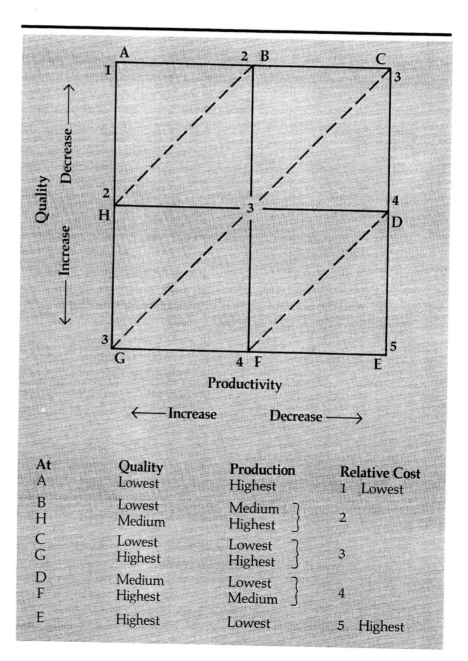

At	Quality	Production	Relative Cost	
A	Lowest	Highest	1	Lowest
B	Lowest	Medium		
H	Medium	Highest	2	
C	Lowest	Lowest		
G	Highest	Highest	3	
D	Medium	Lowest		
F	Highest	Medium	4	
E	Highest	Lowest	5	Highest

Figure 19.2

In Figure 19.2, at corner A, quality is lowest, productivity is highest, and cost is lowest. At opposite corner E, quality is highest, productivity is lowest, and cost is highest. Other combinations fall between these extremes. The cost budget for every item given to the superintendent reflects the estimator's evaluation of the specified quality level, and a productivity ratio is established. In spite of his involuntary commitment to the given budgets, it might be justifiable in some cases for the superintendent to volunteer a higher level of quality than a budget provides, in accordance with longer range company goals (see Chapter 2).

Quality may be divided into 5 classifications:

1. **Items that are not apparent to the eye;** for instance, strength of materials. The sponsor needs certain assurances in this area and usually specifies the type and number of tests he requires. The superintendent requests and receives samples, passes them on to the testing laboratories, receives test results from the labs and forwards them to the sponsor.

2. **Items that are apparent to the eye**, such as material composition, dimensions, color, texture and weight. These can be controversial, since personal taste, preference and opinion are unavoidable, and specs are not always explicit.

3. **Items that are assembled, connected, or mixed together**, such as mechanical components, structural members, or concrete. The integrity of these assemblages, connections and mixes may require proof through various testing techniques. The superintendent must arrange for and cooperate with the specified testing program. He might also arrange for additional testing for his company's own assurance against possible liability.

4. **Aesthetic standards**, such as straightness, plumbness, squareness, uniformity (as in masonry joints). Aesthetic quality has to do with artistry and craftsmanship. It is more difficult to prove compliance with the specs and to obtain the sponsor's approval for aesthetic criteria, since subjectivity plays a major part in its evaluation. For this reason, specs often require the contractor to furnish sample sections of such items as concrete finishes, masonry walls, sprayed texture coating, ceramic tile, precast concrete units, and so forth. This practice is beneficial to both sponsor and contractor, because it establishes solid standards on which both parties can agree.

5. **Items that function**, such as plumbing and mechanical systems, electrical, lawn sprinklers, and elevators. Inspections and tests are performed to assure that these systems work properly.

After the preliminary approval of materials and workmanship, the sponsor requires continuing assurances that the quality is being maintained. Sponsors such as government agencies or large architect and engineer organizations obtain this assurance of ongoing quality through the services of their own on-site inspectors. Tense relationships between superintendents and inspectors sometimes occur, and may lead to disputes, stalemates, and involvement of the Board of Arbitration (see Chapter 5, "Sponsor Relations").

Some sponsors, such as the U. S. Navy Department, have attempted to solve inspection problems by substituting a "disinterested party" for the traditional sponsor-oriented inspector. The *Quality Control Representative* is similar to an independent contractor, having full charge of all testing and inspecting. The quality control representative's salary and all other related costs are covered in the contractor's bid amount and are paid for ultimately by the consumer.

The success of the Quality Control Representative system is controversial. Perhaps the main drawback is the limited pool of available qualified professionals. There might be no perfect solution to the problem of quality control, since this activity falls between the conflicting interests of sponsor and contractor.

In this book, the role of the superintendent is the point of focus; his attitude toward quality is influenced in four ways: the contractor's interest, the sponsor's interest, self interest, and the interest of society.

The contractor's interest is probably his first priority, since satisfying his employer is the best assurance of career longevity. The psychology of team loyalty is also a contributing factor.

The sponsor's interest is strong simply because the progress of the project depends on receiving approvals rather than interferences. Personal conflicts must be kept to a minimum for the same reason.

Self interest might be the strongest incentive to accomplish work of high quality. In addition to justifying his salary, the superintendent establishes or strengthens his reputation with a job well done.

Society's interest is perhaps the most ambiguous, but still a considerable influence as nearly all construction work is for the benefit of human beings.

Considerations not included in the specifications. Not all details are provided in drawings and specs. In their absence, the superintendent must determine heights, locations, and other features that affect human comfort, convenience, and safety.

In Chapter 14—"Time Controls", a system is demonstrated for following through with quality control tests and inspections. At the beginning of a project, the superintendent should make a chart similar to Figure 19.3, designed to expedite ongoing quality control activities after the original submittals have been approved.

A line is drawn on the chart to show the present status of quality control testing and inspections. Everything to the left of the line has been done; everything to the right of the line is yet to be done. Various subcontractors, inspectors, and lab technicians are involved, and the superintendent is able to monitor the action, anticipate the sequences and coach the performances. It is important that tests be made and analyzed as quickly as possible in order to avoid wasting time and money. This approach also prevents the later reconstruction of work that was not validated.

Records of all tests and inspections can have important legal value, perhaps exonerating the contractor in the event of collapses by proving that the materials and workmanship were not at fault.

The data contained in the quality control chart, Figure 19.3, can be programmed into a computer should that technology be available.

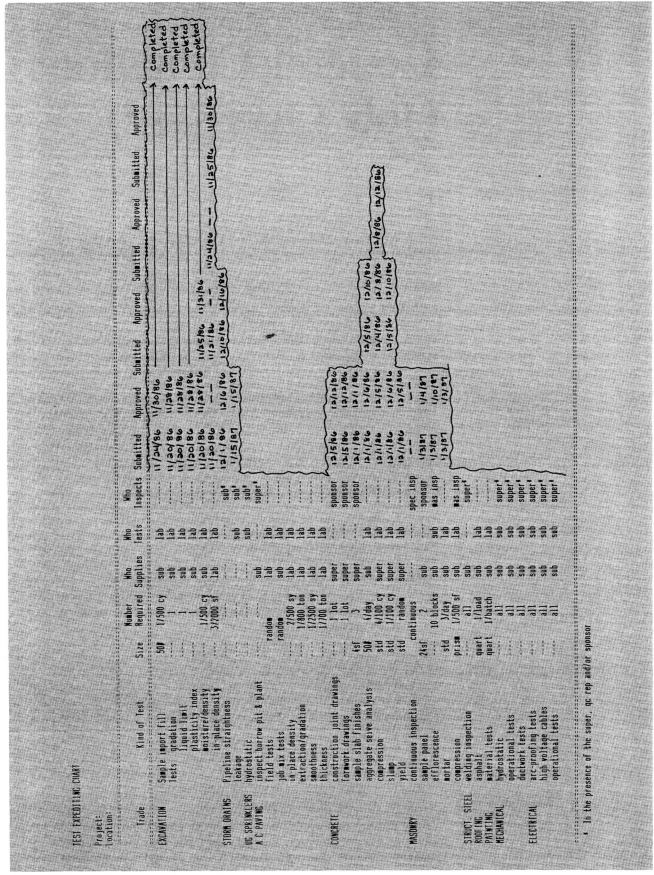

Figure 19.3

Chapter 20

General Conditions (Job Overhead)

In this section, reference is made to the unmarked items in Figure 20.1 (Estimated General Conditions prices). These ongoing general conditions expenditures are sometimes called "job overhead". It is easy for these costs to slide out of control and out of sight, and appear at the end of the project as a huge dent in the profit. Figure 20.2 shows the fixed relationship between General Conditions (GC's) and profit.

It is important that the superintendent have a good grasp of the GC, or "job overhead" costs and at the same time, avoid false economies.

Layout for Structures

Item # 1 in Figure 20.1 starts where the surveyor leaves off; it includes the locations of corners, wall intersections, spread footings and so forth; and it establishes elevations of footings, walls, slabs, and subgrades. The budget is based on a three-man crew for four days which extends to $2,500. Hasty and faulty layout work can lead to expensive repairs and time loss later. By doing some of the layout work himself, the superintendent is better assured of its accuracy. This effort can also help to keep expenditures under control.

In a small project, layout work is done quickly and completely. In a large and complex job, it might stretch through a period of weeks or months. In our present example, the layout will be in two parts: (1) to provide horizontal and vertical control lines for footing excavations, and (2) to provide lines for the concrete formwork. There might also be miscellaneous layout expenses later in the project chargeable to this pay item. This subject is discussed in more detail in Chapter 26—"Layout" and in Chapter 27—"Concrete Footings, Foundations and Slabs".

Monthly Utility Charges

In this example, utilities charges are limited to water and electricity, items # 6b and # 7b. In both cases, the superintendent can only use conservation measures to stay within the budgets of $720 and $840. As in the case of telephone service, underestimating the use of utilities could prove to be false economy.

Temporary Weather Enclosures

Weather enclosures are usually required to keep rainwater from coming through openings which have not yet received doors, windows, and skylights. The cost is always uncertain, and item # 10 is only an allowance.

Costly damage can be done to interior finishes as a result of exposure to moisture. Consequently, the superintendent may keep plywood, plastic sheeting, and other materials on hand to quickly cover any openings when

Means Forms
COST ANALYSIS

PROJECT **Admistration Bldg. & Fire Station**
LOCATION **Rcacho, CA**
TAKE OFF BY **PC**   QUANTITIES BY **PC**   PRICES BY **PC**
CLASSIFICATION   ARCHITECT   EXTENSIONS BY **AT**   **AT**
SHEET NO   ESTIMATE NO   DATE **10/9/87**   CHECKED BY **JM**

Description	Quantity	Unit	Material Unit Cost	Material Total	Labor Unit Cost	Labor Total	Equipment Unit Cost	Equipment Total	Subcontract Unit Cost	Subcontract Total	Total
General Conditions											
1. Layout Structures	4	Days	—	—	625.	2500					2500
*2. Field office	12	Mo.	118.	1416							1416
*3. Storage Trailer	12	Mo.	77.	924							924
*4. Toilets	24	Mo.	73.	1752							1752
*5. Telephone	12	Mo.	150.	1800							1800
6.a. Water Hook-up-Meter	—	L.S.		300		300				1500	2100
b. " monthly Charge	12	Mo.	60.	730							730
*7.a. Electrical Service	—	L.S.		400		500				1268	2168
b. " monthly Charge	12	Mo.	70.	840							840
*8. Security Fence	400	L.F.							9.05	3620	3620
*9. Signs	—	L.S.		100		100				250	450
10. Temp. Weather Closures	—	L.S.		250		800					1050
11. Office Supplies	12	Mo.	35.	420							420
12. Small Tools	12	Mo.	200.	2400							2400
13. Misc. Rental Equip.	12	Mo.	400.	4800							4800
14. Oil, Fuel, Maintenance	12	Mo.	250.	3000							3000
15.a. Clean up, Progressive	12	Mo.	100.	1200		14400					15600
b. ", Final	240	Hrs.	5.	1200		3864					5064
16.a. Superintendent Pay	12	Mo.				44880					44880
b. Foreman	6	Mo.				20208					20208
c. Clerk	12	Mo.				9600					9600
17.a. Barricade at Street	700	L.F.	6.	4200		4200					8400
b. " night Lights	12	Mo.	60.							730	730
18.a. miscellaneous	1	L.S.		1000		420					1420
Sub-Totals				27442		101772				7358	136573

CONSTRUCTION COST QUOTED TO SPONSOR

General Contractor's Construction Cost		Profit
Div. 1	General Conditions	
Div. 2	Site Work	Cost Over-run Eats into Profit
Div. 3	Concrete	
Div. 4	Masonry	
Div. 5	Metals	
Div. 6	Carpentry	Cost Savings Adds to Profit
Div. 7	Waterproofing	
Div. 8	Doors & Windows	
Div. 9	Finishes	
Div. 10	Specialties	
Div. 11	Equipment	
Div. 12	Furnishings	
Div. 13	Special Construction	
Div. 14	Conveying	
Div. 15	Mechanical	
Div. 16	Electrical	

Figure 20.2

weather predictions suggest this necessity (see Chapter 10—"Watching the Weather"). When interior finish work is underway, it may be best to cover openings every weekend regardless of weather predictions. These coverings may also act as a slight deterrent to vandals by screening the interior from view and requiring the overt act of breaking in (see Chapter 7—"Project Security").

Office Supplies

This expense includes an amount for all the needed stationary items as well as depreciation allowances for equipment, such as typewriters, calculators, drafting machines, desks, chairs, a safe, and filing cabinets. Item # 11 is only an allowance. The main reason for recording this expense is for the accurate budgeting of future estimates.

Small Tools

This category consists of company owned tools that need replacing because they are broken, worn out, lost, or stolen. Depreciation is included in this category, along with the cost of sharpening and repairs. The purchase of special new tools is also included. A portion of their cost is chargeable to the current project. Item # 12 is only an allowance.

Miscellaneous Rental Equipment

Cost item # 13, Miscellaneous Equipment, is a catchall for equipment costs not covered in other specific estimated items such as excavation, concrete and carpentry work. The allowance of $4,800 is for equipment which cannot properly be charged to specific work items. This category includes: pumping of rainwater, dust control, general skiploader work, scaffolding and hoisting for subcontractors, and concrete vibrating and accessory equipment not owned by the company. *Oil, fuel, repairs, servicing, and maintenance* are all expenses connected with the use of equipment. Item # 14 is an allowance of $3,000 for these costs. The superintendent can expect a majority of this expense to occur in the first two thirds of the project's time-span and will not be concerned that the monthly costs in that period exceed the budget average of $250 per month. Figure 20.3 shows a probable curve for this item.

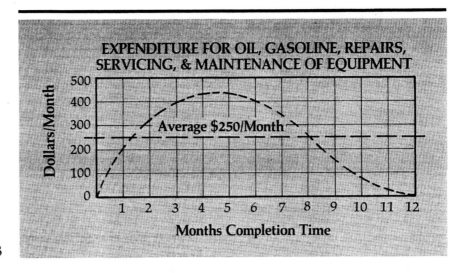

Figure 20.3

Maintaining a clean, orderly building and site increases production and decreases accidents during the construction period. Job cleanup should be accomplished periodically utilizing personnel specifically assigned to the cleanup operation. Many prime contractors assign a cleanup crew one day per week and charge the major subcontractors a portion of the cost. The procedure is usually spelled out in the contract or purchase order between the prime and subcontractor. For purposes of this example, a budget of $16,320 is assigned for cleanup in item 15a. This cost allowance helps to keep other items within the budget. While $1,200 (out of the total $16,320 cleanup budget) is allotted for cleanup equipment, company trucks and skiploaders are usually standing by, available for hauling away construction litter at very little cost.

The $720 budget under the subcontract column is for "dumpster" service (outside trash pickup). This service helps to minimize labor and equipment cleanup costs.

It is fairly easy for the superintendent to keep a running check on the cleanup budget and to remain within it—assuming that the estimated amount is realistic in the first place.

Final Cleanup

This item includes the work required to make the finished project presentable to the sponsor for formal acceptance. Included are:
- Raking, sweeping and hosing down the exterior ground and paved surfaces.
- Cleaning, removing stains and polishing exterior items such as pipe railings, architectural ornamentation, door and window surrounds and fascia.
- Cleaning interior walls, ceilings and floors, removing streaks and stains, waxing floors and polishing hardware.
- Cleaning and polishing bathroom fixtures, accessories, workshop, laboratory, and kitchen equipment.
- Cleaning windows and glass.

In some projects, final cleanup can be extensive and expensive. Item # 15b provides 240 man-hours for this work, which translates to 60 hours for a 4-man crew, or 40 hours for a 6-man crew. Since this is a one-time, end-of-the-job activity, the superintendent can stay within the budget by closely supervising the cleanup work. It should be noted, however, that stinting on final cleanup could be false economy, as a clean job is more saleable than a messy one.

Foreman's Pay

Item # 16b, Foreman's Pay, is an allowance for the probability that the superintendent needs some supervisory assistance. In this example, one of the working foremen might be drafted for assistance at irregular intervals. It is not expected that the superintendent needs a full time assistant for the duration of the project. The superintendent keeps this pay item within budget by either carrying a larger personal work load, or by terminating assistance when the expenditure reaches the budget limit.

Clerk

This category includes a variety of general office duties including:
- Receptionist
- Telephone answering
- Timekeeping
- Processing workers on and off the payroll
- Preparing and/or dispensing paychecks
- Typing
- Filing
- Miscellaneous assistance to the superintendent

Some small, short term construction projects can proceed without the services of a clerk, but in the example, the estimator has provided item # 16c for a full-time, full duration office person to assist the superintendent. The only question of cost control is whether or not a competent clerk can be hired for $800 per month.

Miscellaneous General Conditions

Item # 18a is provided as a blank line item for the recording of costs not anticipated by the estimator. A good example of such an expense is coring and testing of concrete, and earth compaction testing when the results of earlier tests are below acceptable levels (see Chapter 19—"Quality Controls, Inspections, and Tests").

An additional expense might be necessary for dewatering when unexpected underground water is found.

Blank spaces in the cost record sheet for nonbudgeted items serve two purposes: (1) they point out to the estimator certain recurring conditions which should be considered in the bidding of future projects, and (2) they give the superintendent another alternative so that he can avoid charging nonbudgeted costs to the cost record, thereby distorting the budgeted costs.

Chapter 21

Final Accounting (Postmortem)

Immediately following the bidding of a project, a "postmortem" automatically occurs in which the attempt is made to explain what went wrong—or what went right. A similar analysis usually follows the completion of the actual construction work.

Before the project becomes history, the superintendent meets with company management staff for mutual enlightenment regarding unusual details of the project. Typical topics are:

- The actual total (gross) profit compared to the estimated profit.
- The amount of bonuses, if any, earned by foremen and workers.
- Specific cost items that varied greatly over or under the estimate with explanations of the causes.
- Acknowledgment of superior and inferior subcontractors, as judged by attitude and performance.
- Rating of the sponsor and representative.
- A review of mistakes that were made, and proposed methods to avoid their repetition in future projects.
- A review of project photographs taken at different stages and at completion, and selections for framing, publicity, and historical albums.
- Discussion of any particularly effective, innovative construction methods that have been discovered, and their potential benefit to future projects.
- An inventory of leftover materials.
- An accounting of construction equipment returned from the project, and recommendations for repairing, selling or storing it.
- Acknowledgment of outstanding workers and recommendations for retaining or promoting them.
- Backcharging and other matters relating to unresolved final subcontractor payments.
- Discussion of ways to improve field office routine work, such as the use of forms and computers.

Part III

Superintending the Trades

Chapter 22

Stress Calculations, Precision and Waste Factors

Generally, the superintendent's responsibility for performing stress calculations is limited to items incidental to construction work such as scaffolding, concrete formwork, and shoring. In many cases, judgment based on experience is an acceptable substitute for calculations. Occasionally, however, formal proof of the structural adequacy of materials, sizes of members, bracing and connections is a spec requirement. The specs might specifically call for the services of a professional engineer, or the superintendent might decide—for precautionary reasons—to hire an engineer to do spot calculations. In most cases, pre-engineered tables of structural data are sufficient for the design of formwork and shoring. Many material suppliers furnish stress information with their products. Still, it is good practice for the superintendent to make as many stress calculations as time permits and to keep a record of them for two reasons: (1) improvement of his own ability to do calculations, and (2) proof, should any be required, of his carefulness.

Precision involves (1) accuracy of measurements and computations, and (2) the degree of perfection in work such as lines, levels, and surfaces. It is often impractical or even impossible in construction to accomplish an absolute standard of workmanship; therefore, certain tolerances for deviations are allowed both by specifications and by custom. Let us consider measurements and computations first. Every measurement is composed of a number, *an estimate of its uncertainty*, and a unit. Some causes of uncertainty are faulty measuring instruments, human error, temperature, friction, imperfect tools, defective materials, and accidents.

The relative error formula is a method for identifying and setting acceptable limits to deviations from drawings and specs. It is the ratio of the measured error to the specified (standard), or

$$RE = \frac{error}{standard}.$$

For example, what is the RE of an angle specified to be 60 degrees, but measured 59 degrees—15'? The error $= 45'$, or .75 degrees; $RE = .75/60 = .0125$

Materials differ in RE due to their varied physical characteristics and workability; but designers occasionally specify tighter RE's than normal, requiring exceptional care and consequently, greater time and cost. At the start of a project, the superintendent might read through the specs and list all

Acceptable Tolerances
(Example Only)

	Normal Tolerance + or −	Normal Re	Specified Re
Earthwork-typical surfaces	1/10′	.10	.10
compaction, relative	5%	.05	NS
Survey lines, angles, elevations	$\frac{1}{2}''$ in 100′	.0001	NS
Gravity pipe lines-leakage	{ 500 gals per 1″ of diam per day per mile }	.095	.095
grade line	2″/100′	.002	NS
A.C. paving-base course	1/10′	.10	.10
surface	3/8″/10′	.038	(.002)*
thickness	3/8″/4″	.094	NS
Parking striping-alignment	2″/20′	.009	NS
stripe width	$\frac{1}{2}''$/6′	.038	NS
Concrete formwork-horiz. & vert.	1/8″/10′	.013	(.020)*
Concrete-length	1/8″/10′	.013	NS
cross section-to 6″	1/8″	.167	NS
6″ to 18″	3/16″	.010	NS
18″ to 36″	1/4″	.008	NS
over 36″	3/8″	.010	.010
camber	1/16″/10′	.001	(.002)*
slab surface	1/4″/10′	.002	.002
slab thickness	3/8″/4″	.094	NS
slab slope to drain	3/8″/20′	.002	NS
Masonry-vertical	1/4″/10′	.002	.002
horizontal	1/2″/20′	.002	.002
thickness	1/4″/8″	.031	.031

*⬭ circled numbers vary from normal
NS = not specified

Figure 22.1

the required RE's. The list will then help him avoid costly quality errors. Figure 22.1 is an example of a list with one column of assumed normal RE's and another column of RE's as required by specs. These are examples of maximum deviations from ideal standards and are generally plus or minus, as illustrated in Figures 22.2 and 22.3.

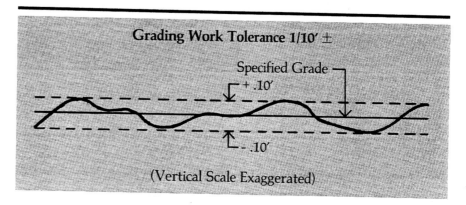

Figure 22.2

The superintendent notes the items that are specified at normal RE levels and those that vary from normal RE's. The latter are circled so that special attention will be drawn to them. Other items are not specified at all (NS). The superintendent assumes that normal RE's will be accepted by the sponsor's representative.

Because expansion and contraction can be damaging, the superintendent should give special attention to expansion joints, their dimensions, locations and spacings. Joints need to be wide enough to allow for material expansion, but not so wide as to cause structural weakness or water damage. Expansion joints will be discussed under the section on concrete work.

Some *waste* in the use of materials is unavoidable. Cost budgets include allowances for waste in the estimated quantities, but the superintendent should judge and compute these values for himself.

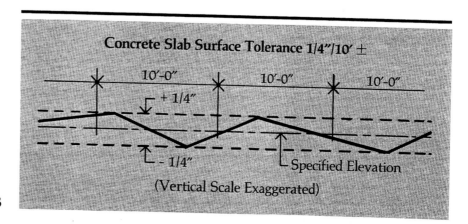

Figure 22.3

Waste is the quantity of loose material that must be purchased in excess of that required for net, in-place. It is sometimes convenient to use the term "waste" even when the excess over net quantity is due partially or entirely to such causes as compaction, or loss through vandalism or theft. Figure 22.4 tabulates some causes of waste. The rest of this section covers in detail each of the causes listed in Figure 22.4.

Causes of Waste	Examples
Shrinkage	compacted earthen materials
Hardening, spillage, spoilage-	concrete; caulking
Breakage	glass; lumber
Wear and tear	forming materials; scaffolding
Cutting	any materials
Dressing	lumber
Overlapping	plastic sheeting, shingles
Over-ordering quantities	any materials
Theft or vandalism	any materials
Mistakes, measuring, etc.	any materials
Damage, rain, etc.	any materials

Figure 22.4

Shrinkage allowances for various soils may be approximated by the use of a table, such as that shown in Figure 22.5. The net cubic yards compacted in place (bank cubic yards) is increased to loose cubic yards by adding the appropriate percentage; example: how many loose cubic yards of common earth must be ordered to backfill a trench requiring 180 bank cubic yards? Answer: 180 x 1.25 = 225 cubic yards.

Hardening, Spillage and Spoilage

This type of waste occurs with fluid, plastic, or loose materials. Materials that spill or blow away or become contaminated may be uneconomical as they are impossible to retreive. Wet concrete that hardens before it is placed may have to be disposed of (wasted). Caulking, paint, asphalt, etc., might harden in containers. Spoilage might occur through wet or dry rot, water staining, or rodent or termite action. Materials kept in storage for a long time, no matter how carefully protected, might deteriorate to the point that a portion becomes waste. Any allowance for this kind of waste can only be classified as miscellaneous. Climate and other factors such as project locations might affect the percentage of waste. Unless conditions are unusual, a small percentage (such as 1%) should be an adequate allowance at the time of estimating and/or ordering.

Breakage

Many kinds of materials are subject to breakage by accidents such as falling, being run over by trucks, or as the result of excessive loading and stacking. To be realistic, the superintendent should make an allowance for this kind of waste of, say, 1%.

Wear and Tear

"Wear and tear" can damage materials to a point that replacement becomes necessary. An aggregate subbase may be partially crushed or compacted into the subgrade if it is exposed to heavy traffic, and additional material may be needed to reestablish the grade level. Concrete floor slabs used for precast panel casting surfaces can be damaged by the use of inferior bond breaking compound, and may require patching.

Concrete forming materials are the most conspicuous example of the effects of wear and tear. Materials such as metal, fiberglass, or plastic-coated plywood have high resistance to wear and tear; they provide the

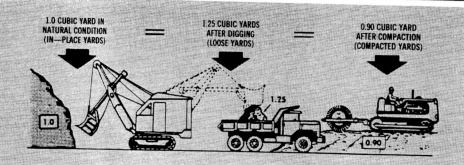

| 1.0 CUBIC YARD IN NATURAL CONDITION (IN—PLACE YARDS) | = | 1.25 CUBIC YARDS AFTER DIGGING (LOOSE YARDS) | = | 0.90 CUBIC YARD AFTER COMPACTION (COMPACTED YARDS) |

Approximate Material Characteristics*

Material	Loose (lb/cu yd)	Bank (lb/cu yd)	Swell (%)	Load Factor
Clay, dry	2,100	2,650	26	0.79
Clay, wet	2,700	3,575	32	0.76
Clay and gravel, dry	2,400	2,800	17	0.85
Clay and gravel, wet	2,600	3,100	17	0.85
Earth, dry	2,215	2,850	29	0.78
Earth, moist	2,410	3,080	28	0.78
Earth, wet	2,750	3,380	23	0.81
Gravel, dry	2,780	3,140	13	0.88
Gravel, wet	3,090	3,620	17	0.85
Sand, dry	2,600	2,920	12	0.89
Sand, wet	3,100	3,520	13	0.88
Sand and gravel, dry	2,900	3,250	12	0.89
Sand and gravel, wet	3,400	3,750	10	0.91

*Exact values will vary with grain size, moisture content, compaction, etc. Test to determine exact values for specific soils.

Typical Soil Volume Conversion Factors

Soil Type	Initial Soil Condition	Bank	Converted to: Loose	Converted to: Compacted
Clay	Bank	1.00	1.27	0.90
	Loose	0.79	1.00	0.71
	Compacted	1.11	1.41	1.00
Common earth	Bank	1.00	1.25	0.90
	Loose	0.80	1.00	0.72
	Compacted	1.11	1.39	1.00
Rock (blasted)	Bank	1.00	1.50	1.30
	Loose	0.67	1.00	0.87
	Compacted	0.77	1.15	1.00
Sand	Bank	1.00	1.12	0.95
	Loose	0.89	1.00	0.85
	Compacted	1.05	1.18	1.00

Figure 22.5

superintendent a choice between reusing cheaper materials a few times, or more expensive products more times. An evaluation includes the cost of both materials and labor. Following is an example evaluated for (1) inexpensive plywood, (2) metal paneling, and (3) plastic-coated plywood. The cost of hardware is omitted, since it is roughly the same for each method.

Form 24,000 square feet of retaining walls.

Method 1-Plywood (4 uses)

	Quantity	Unit Prices l	Unit Prices m	Extensions L	Extensions M	Extensions T
Lumber	10,000 bf	—	.40	—	4,000	4,000
Plywood	3,500 sf	—	.40	—	1,400	1,400
Waste 30%	4,050 bf	—	.40	—	1,620	1,620
Labor	24,000 sf	2.90	—	69,600	—	69,600
Subtotal				69,600	7,020	76,620
Less Salvage	13,500 bf	—	.05	—	(675)	(675)
Total 75,945/24,000 = 3.16/sf				69,000	6,345	75,945

Method 2-Metal paneling (8 uses)

	Quantity	l	m	L	M	T
Lumber	6,000 bf	—	.40	—	2,400	2,400
Panels, rented	3,000 sf	—	2.25	—	6,750	6,750
Waste 15%	900 bf	—	.40	—	360	360
Labor	24,000 sf	2.60	—	62,400	—	62,400
Subtotal				62,400	9,510	71,910
Less Salvage				—	—	—
Total 71,910/24000 = 3.00/sf				62,400	9,510	71,910

Method 3-Plastic-coated plywood (8 uses)

	Quantity	l	m	L	M	T
Lumber	6,000 bf	—	.40	—	2,400	2,400
Plastic plywood	3,000 sf	—	1.20	—	3,600	3,600
Waste 20%	1,800 bf	—	.40	—	720	720
Labor	24,000 sf	3.00	—	72,000	—	72,000
Subtotal				72,000	6,720	78,720
Less Salvage				—	(2,700)	(2,700)
Total 76,020/24,000 = 3.17/sf				72,000	4,020	76,020

In this example there are five variables: (1) original material cost, (2) number of reuses, (3) labor, (4) waste, and (5) salvage value of materials. Rented metal forming appears to be the most economical method because in spite of the highest material cost and the least salvage value, labor and waste are substantially reduced. However, metal forming is not always the best choice. In different kinds of forming jobs, first one and then another material will have the advantage; and cost is not always the primary consideration. Sometimes the desired appearance of the concrete surface will dictate the choice of forming material.

Cutting waste is fairly constant among different materials and usages. Figure 22.6 is a sampling of the various percentages for cutting waste allowances.

Since material lists itemize pieces in standard lengths, they automatically allow for a portion of the cutting waste; therefore, the minimum percentages in Figure 22.6 are usually sufficient.

Cutting Waste

Material	Percentage From	To
Forming lumber	4%	10%
Forming plywood	6%	11%
Reinforcing steel	5%	9%
Framing lumber-studding	2%	4%
wall blocking	10%	15%
floor joists	3%	5%
roof joists	4%	6%
beams	5%	8%
columns	4%	7%
Plywood sheathing	4%	8%
Plain boarding, 1 × 6 perpendicular	10%	13%
diagonal	15%	20%
Plain decking, 2 × 6 perpendicular	8%	11%
diagonal	12%	17%
Gypsum board	11%	16%
Plywood wall paneling	13%	20%

Figure 22.6

Since material lists itemize pieces in standard lengths, they automatically allow for a portion of the cutting waste; therefore, the minimum percentages in Figure 22.6 are usually sufficient.

Dressing waste occurs in special placement cases, such as tight boarding, siding and decking. Figure 22.7 is a sampling of percentages for typical dressed lumber waste.

Another example of dressing waste is the milling down of standard sizes to net dimensions, such as $3'' \times 4''$ to $2\frac{1}{2}'' \times 4''$. In such cases, 3×4's and milling costs are figured, and waste as such is ignored.

Overlapping waste occurs with materials such as plastic membranes, building paper and lap-siding. The percentage of waste depends upon the widths or lengths of pieces and the amount of overlap. Figure 22.8 is a sampling of typical overlapping waste.

Over-ordering of quantities can cause waste, if the materials are not returnable, or if it is not economical to return them. This kind of waste is common because it is almost impossible to order the exact quantity—no more nor less than is needed for in-place work. It is usually more economical to purchase a bit extra than to buy short. Obtaining a slight safety margin is not considered over-ordering. When it does happen, the cost-conscious superintendent may attempt one or more of the following methods to dispose of the excess:

- Return it to the supplier.
- Use it on the site; for instance, excess concrete can make wider or deeper footings or slabs.
- Store in the company's general yard or warehouse. If this method is used, ask the bookkeeper to charge the cost to inventory rather than to the job.
- Find a buyer. Selling, even at a discount, can minimize the loss.

Percentage Waste in Dressed Lumber

		Percentage Waste
Boarding, plain	1 × 6	10%
	1 × 8	7%
	2 × 6	10%
	2 × 8	7%
T&G	1 × 6	20%
	1 × 8	14%
	2 × 6	20%
	2 × 8	14%

Figure 22.7

Theft or vandalism waste is not usually anticipated or provided for in the budgets. Losses of this kind may be covered by the company's insurance. In any event, such costs should not be entered into the project cost record as normal charges.

Mistakes in measuring and calculating are neither anticipated nor allowed for in the budget—but they do happen. Mistakes on the short side are a form of waste, since reordering and delays result. Excess material might also create a cost problem, as discussed above in the section on over-ordering.

Damage due to rain and temperature extremes can also cause waste. Company insurance might cover such losses under certain circumstances. However, additional materials must often be purchased at the contractor's expense to replace unusable damaged stock.

Completely preventing material damage may be impossible, but some degree of protection is needed. See Chapter 6 for discussion of protective measures.

As a general rule when calculating material quantities, a token allowance of, say, 1/2 of 1% (.5%) could be included for damage. Figure 22.9 is an example of a material list. It shows various materials as well as allowances for the different classes of waste itemized in this section.

Overlapping Waste

		Percentage Waste
Plastic membrane —	10′ wide, 6″ lap	5%
	12′ wide, 6″ lap	4%
Building paper	4′ wide, 2″ lap	$4\frac{1}{2}$%
	6′ wide, 2″ lap	3%
Siding	6″ wide, 1″ lap	19%
	8″ wide, 1″ lap	16%

Figure 22.8

Example Quantities with Waste Allowances

Material	Net Quantity	Kind of Waste	% Added for Waste	Total Quantity
Decomposed granite	680 tons	shrinkage	25%	850
		wear/tear	1%	7
				857 tons
Topsoil	215 cy	shrinkage	5%	226
		wear/tear	1%	2
				228 cy
Forming materials	10,000 bf	wear/tear	30%	13,000
		breakage	1%	100
		cutting	5%	500
		damage	$\frac{1}{2}$%	50
				13,650 bf
Concrete	330 cy	hardening	1%	333
		spillage	1%	4
		wear/tear	1%	4
		mistakes	1%	4
		over-order	1%	4
				349 cy
Plastic sheeting	16,000 sf	overlapping	5%	16,800
		damage	$\frac{1}{2}$%	80
				16,880 sf
Lumber, framing	88,000 bf	cutting	4%	91,520
		breaking, theft, damage	3%	2,640
				94,160 bf
Lumber, 2 × 6 T&G	17,000 bf	cutting	8%	18,360
		dressing	20%	3,400
		theft, damage	2%	340
				22,100 bf

Figure 22.9

Chapter 23

Detailing and As-Built Drawings

Perhaps one third or more of a superintendent's time is spent referring to the drawings and specs. The information contained in them is usually sufficient for the purposes of estimators and bidders; it is seldom completely sufficient for the superintendent's purposes. Additional detailing, dimensioning and separate working (shop) drawings are often necessary. In some cases, the specs call for extra detailing to be performed by the contractor, subject to architect and engineer approval. In other cases, the superintendent chooses, on his own initiative, to do extra detailing for an obvious need (formwork for concrete, for example). Detailing applies to all parts of a construction project, to less significant as well as major items.

Large scale or complex projects may warrant the employment of a detailer. Projects of medium complexity may require the foreman to assist the superintendent with detailing work. Basic drafting skills are used to measure, plot, and illustrate.

All detailing should be retained and made a part of the official project documents. To a limited extent, this information may be inked onto the original drawings to serve as "as-built" drawings.

"As-built" drawings are of great value to the sponsor for future maintenance and alterations, when the exact locations of such items as footings and pipes are needed. The immediate value of these drawings to the superintendent is to explain problems, make decisions, and receive approvals, thus avoiding possible disputes.

Detailing usually begins and ends with site work. It begins with site work because clearing and earthwork must be performed before any construction can be started. It ends with site work because certain items, such as walks, curbs, pavements, and landscaping, can only be done after the general construction work is completed.

The detailing in a typical project might include the following:

1. Locating and recording existing underground utilities, some of which have been discovered in the course of placing new utilities. Figure 23.1 is an example of the way this kind of information might be plotted on as-built drawings.

   Existing utilities are located by measured reference to one or more fixed objects above ground. The depth is generally recorded as the distance from a known, fixed elevation point. This information might be valuable to the architect and engineer, the sponsor, and to municipal engineers in regard to future projects.

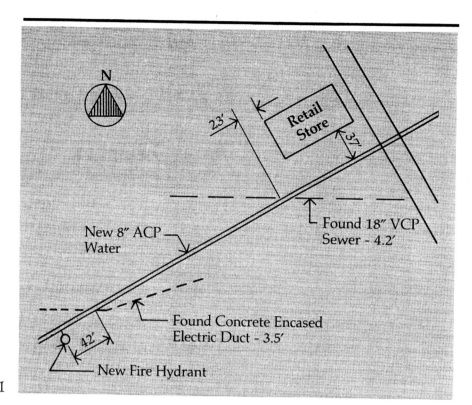

Figure 23.1

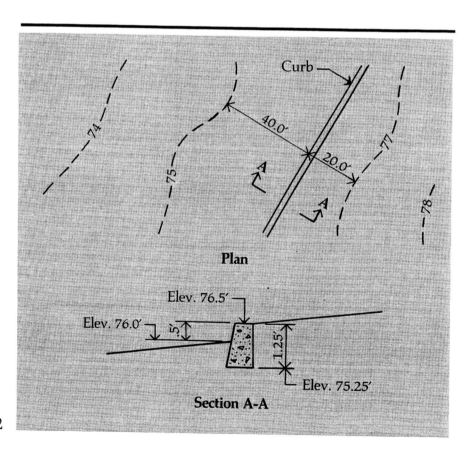

Figure 23.2

2. Interpolating grade elevations at special points not given on the drawings. Most of this kind of calculating falls under layout work (see Chapter 26). Certain critical points might be determined by the superintendent (or detailer) and recorded on the as-builts. Figure 23.2 is an example of an interpolation to locate a curb elevation. Note that the choice was made to interpolate from the contours to the right of the curb. If the contours to the left had been used, different elevations might have resulted.

A few of these strategic control points can be used to easily locate several elevations. In this way, the superintendent is able to check the accuracy of a subcontractor's excavation and layout work. The work of other trades may also be checked in this manner. Figure 23.3 is an example of the detailing required to locate the top elevation for a catch basin (area drain).

3. Establishing various construction and control joints in concrete slabs, walls, and structural members. The locations of some joints is mandatory, as shown in the drawings; others are optional and left to the superintendent's discretion. Figure 23.4 is an example of a floor slab with various joints and their spacings indicated, ready for the sponsor's approval and work crew guidance. Superstructure construction joints can be ambiguous in the drawings, and may be left to the superintendent's discretion. The locations of these joints should be clearly determined to guide the placing of reinforcing steel and the design of forms. Figure 23.5 is an example of superstructure detailing.

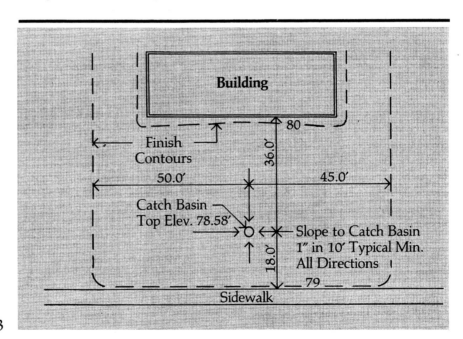

Figure 23.3

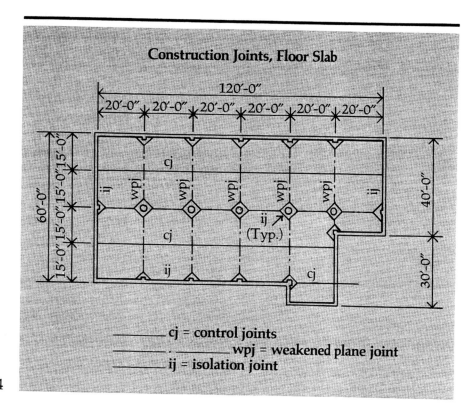

Figure 23.4

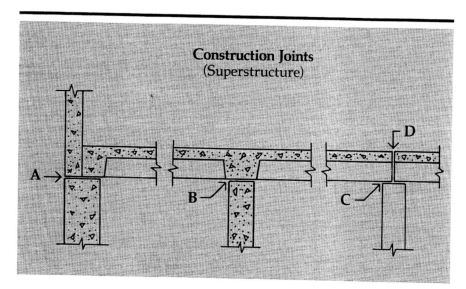

Figure 23.5

Chapter 24

Demolition, Clearing, and Grubbing

In preparation for construction work, a site usually requires *clearing* of trees and vegetation, *demolition* of existing structures, and removal of other interfering objects. This process is followed by *earthmoving*, which includes cutting, filling, compacting, sloping, and fine grading. Finally, *bulk excavating* with heavy equipment may be required for basements, pits, embankments, tunnels, ditches, ponds, bridge abutments, and the removal of unstable subgrade materials. Structural excavation (usually done by *light* equipment, or by hand labor) is not included in this section; it is described in Chapter 27, "Footings, Foundations, and Slabs".

Clearing, earthmoving, and mass excavation are usually sublet to earthmoving specialists. Even so, the superintendent must keep in close and constant communication with the subcontractor to ensure the following: that the correct objects are cleared from the site, the grades are made to the proper slopes and elevations, the watering and compacting is optimum, the fill materials comply with spec requirements, fine grading is done within acceptable tolerances, and mass excavations are made in the correct locations and to the proper dimensions.

Clearing and grubbing includes such items as trees, brush, and tree stumps not ordinarily taken out by demolition specialists. Removal of interfering objects might involve dismantling and hauling away paving, curbs, building foundations and slabs, fences, culverts, and headwalls. Catch basins, manholes, and underground piping may also have to be abandoned. When such jobs are clearly let to a subcontractor, the superintendent may have only to identify the specific items. However, if these tasks are excluded by the grading subcontractor, they remain the direct responsibility of the superintendent.

Clearing of brush: This work usually requires a combination of equipment, such as dozers and loaders, in addition to hand labor. The main problem is disposal. Well written specs will instruct the superintendent in this matter. Otherwise, he may choose between: (1) burning (if permitted, this is the quickest and most economical method); (2) pushing off-site and abandoning; (3) chipping and then storing for use as mulch, or hauling to disposal areas; and (4) loading on trucks and hauling to disposal areas. Disposal of brush, as well as most other cleared debris, is an environmental issue; the contractor might be liable for illegal disposal (see Chapter 8 for further discussion of environmental considerations).

Tree Removal: This work is ideally sublet to tree specialists. There is no sharp distinction between brush and trees; therefore, brush clearing by dozer may automatically take out small trees. The remaining root systems might have to be cut out in a separate stripping operation. Since stump and root removal can be a bigger job than cutting the trees, it may sometimes be easier to push the tree over with heavy equipment, or pull it over by cable. The stump is then cut free from the trunk, loaded on a truck and hauled away. If (for reasons of safety or convenience) uprooting is not the chosen method, saw cutting at a height of four or more feet above ground will provide leverage for uprooting the stump by backhoe, dozer, or cable. In either case, the smaller limbs may then be cut away from the trunk and disposed of in the same manner as brush (described above). The larger limbs and trunk may be: (1) cut into lengths that can be loaded on trucks for disposal, (2) dragged by tractor with logging chains to a nearby, acceptable off-site location and sold (with or without further cutting), or (3) cut, split, and stacked off-site as usable or saleable firewood.

Removal of boulders: This operation can be carried out using various methods. The choice may be governed by contract specifications. Burying or blasting are among the alternatives, which also include using the material for ornamental purposes, or as rip-rap in designated locations. Boulders may also be hauled off-site. Depending upon spec requirements, the following list contains other possibilities for boulder removal:

- Small and medium size boulders may be:
  a. bulldozed or trucked, or carried in loader buckets to low areas where they will be covered over with fill soil.
  b. placed on slopes for rip-rap.
- Large boulders may be broken up by blasting, backhoe with demolition hammer, jackhammer, or plug and feather, then disposed of in the manner of small and medium boulders.
- All sizes might be:
  a. reduced by crushing to an aggregate which could be used for on-site paving or base course.
  b. loaded on trucks and removed from the job site.

Removal of Pavement

Whether asphaltic concrete or other, pavement removal often begins with saw cutting or jack hammering with a wide bit to provide neat and minimum patching. Even if saw cutting is not required by the specs, it might be a worthwhile expenditure for the contractor. If the contractor does not own cutting equipment, the superintendent may either rent a machine (with or without operator) or obtain unit price quotations and sublet the work. In any case, the superintendent must lay out and mark the pavement to guide the cutting machine.

The method for breaking out depends on the type of pavement, the quantity, and the conditions for disposal. Typical methods are:

- A small quantity may be broken up more economically by hand-held pneumatic hammers rather than heavy equipment (if heavy equipment is not already available on the job site). The broken pieces can then be disposed of as debris either by hand or with a small loader.
- A moderate quantity may be broken out by backhoe, dozer, or bucker loader.
- A large quantity may be broken into manageable size pieces by a heavy pavement breaking machine.

Typical methods for disposal include:
- If the broken material can be used on or near the site (for rip-rap or burial under fills) it may be moved by a bucket loader of appropriate size.
- It can be loaded on trucks and hauled to the nearest dump site.
- Asphalt pavement might be sold to a recycling plant. In some projects it may be recycled in a portable on-site plant for use as new pavement or as base course material. It may be feasible to crush broken concrete and use the resulting aggregate for base courses or for low strength concrete.

Removal of Concrete Curbs, Foundations and Floor Slabs

This job is often done most efficiently using a combination of hand-held pneumatic hammers, backhoe with demolition hammer, loader and trucks. One or more workers with steel cutting tools might be required for cutting reinforcing steel. A small dozer may also be needed to remove earth around footings and retaining walls.

Abandoning of Manholes, Buried Piping, and other Miscellaneous Items

Because this work entails a lot of excavation, it is often incorporated into the earthmoving operation. It is relatively easy for a heavy bulldozer, while doing mass excavation, to hook under such buried items and pull them clear for breaking, loading, and disposal.

It is not easy to compare items on the estimating department's cost budget to actual, individual costs for clearing and removal work, since this work is done on a piece-by-piece basis. Workers and equipment clear away items rapidly and randomly. When this phase of the work is complete, it is easier to compare the total actual cost to the total estimated cost. Nevertheless, the superintendent is expected to make the best possible attempt at separate item cost accounting.

Chapter 25

Earthmoving and Bulk Excavation

Earthmoving, also called "grading" or "rough excavation", usually begins with stripping the surface of materials that are unacceptable for use in embankments. This category includes: grass, weeds, leaves, mulch, roots, topsoil, and loose rocks. The exact desirable depth of removal is not always easily determined. Consequently, specs may stipulate minimum depths (such as 4") for stripping. The cost conscious superintendent is reluctant to overstrip, because this overstripped material costs extra—both to move initially and to put back in place. Should the material below the stipulated stripping depth be unacceptable, the superintendent may wish to negotiate with the sponsor for extra compensation.

Who determines "acceptability"?—usually the sponsor's representative or the quality control representative with or without the aid of laboratory soil tests. Some specs call for the services of a professional soils engineer whose decisions regarding stripping and all other earthwork matters are final. Stripping is usually approached as a separate operation, though an exception might be made when the total quantity of cut exceeds the quantity of fill needed. The extra amount is waste or spoil. It might be more economical to excavate and dispose of a portion of the excess material (including the stripping layer) in one integral operation, as in Figure 25.1.

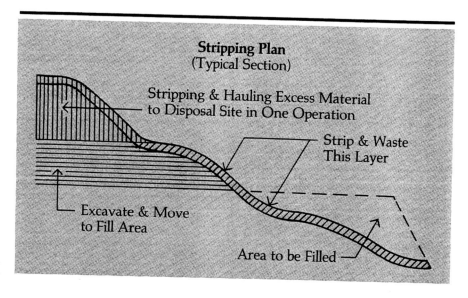

Figure 25.1

Stripping Plan
(Typical Section)

Stripping & Hauling Excess Material to Disposal Site in One Operation

Strip & Waste This Layer

Excavate & Move to Fill Area

Area to be Filled

Stripping on rough terrain is most efficiently done by bucket loader (assisted sometimes by a dozer). In less accessible spots, the backhoe is useful. On relatively large, flat areas, a grader can scrape the surface layer into windrows for loading on trucks. In unusually large areas, rubber tired scrapers can very quickly accomplish the stripping work.

As a general rule, it is better to have the survey staking follow the stripping operation, as this prevents the upsetting or losing of stakes. Surveying is discussed in Chapter 26.

Grading: This follows the stripping work if it is done by a subcontractor. It involves the superintendent with the following concerns:

- That proper (specified) materials are used in the fill areas.
- That minimum compaction requirements are met (such as 90%, 95%, etc.).
- That the dimensions and elevations of cuts and fills are correct.
- That the slopes and contours are suitable for proper drainage.
- That finish grades are accurate to within accepted standards of tolerance (typically $+$ or $-1/10'$).
- That all earthwork is complete in minimum time and within the previously planned progress schedule (see Chapter 14, "Time Controls").

These issues are of equal concern to the superintendent and the subcontractor, and close collaboration and cooperation are to their mutual advantage. Figure 25.2 is an example of an ambitious earthmoving project programmed for maximum production. The average truck cycle time through city streets is found to be 16 minutes, including loading, travel time, dumping, and getting ready for reloading.

A five cubic yard loader places two heaping buckets (10 cubic yards) of loose fill in one truck in 2.67 minutes. The maximum number of trucks that can be used is

$$\frac{16 \text{ (minutes travel time per truck)}}{2.67 \text{ (minutes loading time per truck)}} = 6 \text{ trucks}$$

At the job site, two dozers provide enough excavated material to keep the loader working full time. At the disposal site, two dozers spread the fill as fast as it is dumped. No compaction is required, but a water truck is needed to control dust. The trucks maintain an average time distance between them of 2.67 minutes. The total quantity moved per hour is

$$\frac{60 \text{ (minutes per hour)}}{16 \text{ (minutes travel time per truck)}}$$

$\times$ 6 trucks x 10 (cubic yards of fill/truck/per trip) = 225 cubic yards.

The total time to move 24,750 cubic yards is

$$\frac{24,750 \text{ (total cubic yards)}}{225 \text{ (cubic yards moved in one hour)}} = 110 \text{ hours} = 13.75 \text{ days}$$

The total cost is: \$550/hr. x 110 hrs. (total time) = \$60,500

The unit cost is:

$$\frac{\$60,500 \text{ (total cost)}}{24,750 \text{ (total cubic yards}} = \$2.44/\text{cubic yard}$$

The time and cost might be decreased if larger equipment could be used—but the street traffic will not accommodate larger trucks. The superintendent should realize that this plan ignores the possibility of equipment breakdowns, and should insist that the equipment supplier have available standby equipment. The plan also assumes uniform material throughout; should harder earth be encountered, the numbers and types of equipment would change.

Excavation, Disposed Off Site By Truck

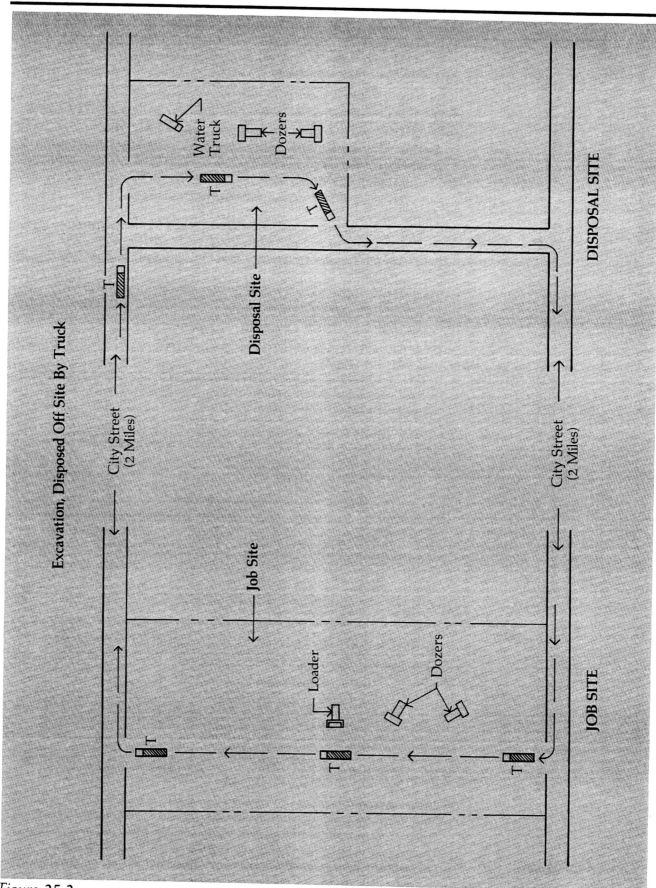

Figure 25.2

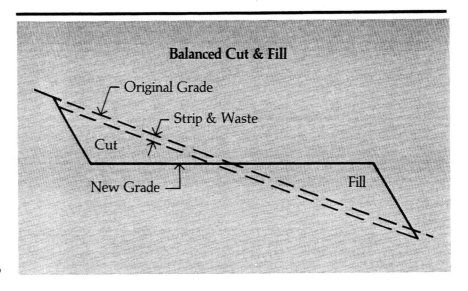

Balanced Cut & Fill

Original Grade

Strip & Waste

Cut

New Grade

Fill

Figure 25.3

Some projects are designed by the architect and engineer so that cuts and fills (after stripping) will balance, as in Figure 25.3. Most projects do not balance, as in Figure 25.4.

Disposing of excess cut material can be more expensive than cutting and filling on-site. The desirable course of action would be selling the excess fill or finding a nearby disposal site.

Procuring fill material can be even more expensive than disposing of any excess, since it might have to be purchased at a premium price, or hauled from a distant borrow pit.

If the superintendent can interest the sponsor (by offering a credit) in balancing cuts and fills by adjusting finish grade elevations, the benefits can be mutual and substantial.

The project specs usually outline the requirements for filling work. Figure 25.5 shows some of the typical features.

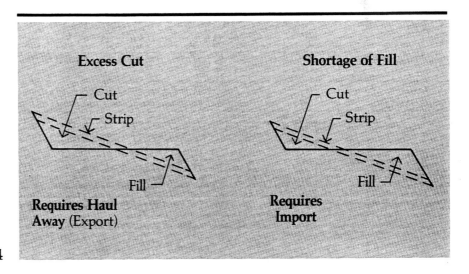

Excess Cut

Cut

Strip

Fill

Requires Haul Away (Export)

Shortage of Fill

Cut

Strip

Fill

Requires Import

Figure 25.4

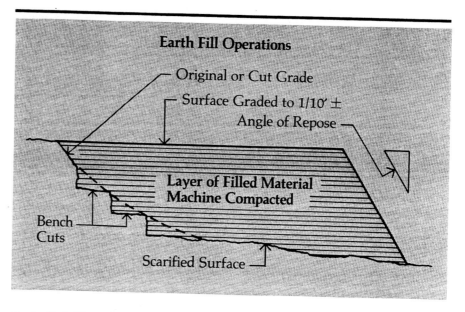

Figure 25.5

Controlled fill is placed in layers 6" or 8" deep, watered to an optimum moisture content and compacted to a specified density, as determined by laboratory tests. Benches are cut into the original or cut slopes against which the fill is to be placed to prevent faulting or landsliding. Ground surfaces on which fill is to be placed should first be scarified to a depth of 6" or more to achieve integration between the original soil and the new fill material. Banks are sloped to specified angles (such as two vertical to one horizontal) in order to avoid sliding. The finished surface is graded to the specified contours and to an accuracy of within 1/10 of a foot. A straight board 10' to 20' long can be used to test the accuracy of the finish grade; an engineer's level and rod may also be used over larger areas, but often a mere visual inspection is sufficient.

Bulk Excavation

Bulk excavation or structural excavation is of special interest to the superintendent. This work is often not included in earthmoving or grading subcontracts. The superintendent then becomes personally responsible for the planning and performance of these excavations.

Excavations for foundations, manholes, sewage lift stations, electrical vaults, and elevators can present various and sometimes unexpected problems. The decisions required by these situations may involve:

- Pavement sawcutting, removing and patching
- Working room and space to store backfilling material
- The need for shoring
- Underground water and the need to dewater
- Disposal of excess soil
- Backfilling around structures

Figure 25.6 shows two ways to excavate a pit for an underground structure.

Following is a comparison of methods:

Method A—*Disadvantages*

- Requires a larger perimeter working area
- Involves a larger area of pavement removal
- Requires a larger volume of excavation and backfill
- Requires a larger area of pavement patching
- Presents a greater probability of cave-ins

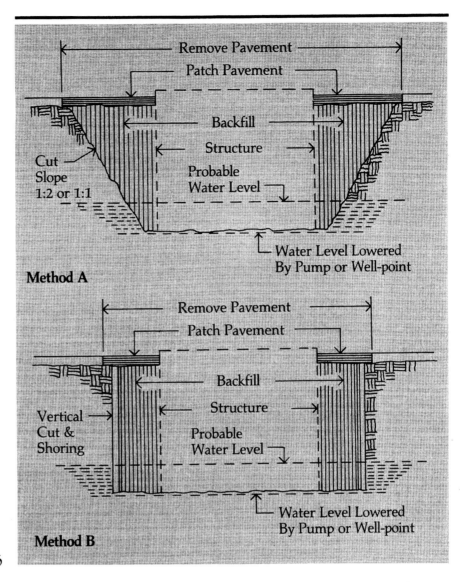

Figure 25.6

Advantages
- Requires a shorter construction time
- Involves no shoring expense

Method B—*Disadvantages*
- Has a slower completion time due to shoring
- Involves expense of shoring

Advantages
- Requires smaller perimeter working area
- Involves smaller volume pavement removal and patching
- Requires a smaller volume excavation and backfill
- Is less prone to cave-ins

The superintendent should be familiar with the swell and shrinkage characteristics of various earthen materials. Figure 25.7 shows how swell applies to excavation work and how shrinkage applies to filling.

Often the material excavated from a pit will not fill it again. If it is well compacted, additional material may have to be imported. The reason is that artificially compacted earth can be more dense (fewer voids) than naturally settled earth. Consequently, it is good practice to order three to five percent

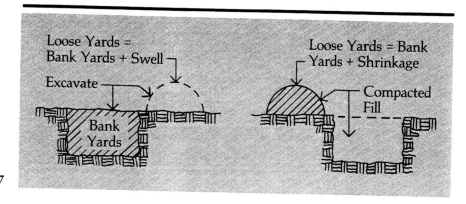

Loose Yards =
Bank Yards + Swell

Excavate

Bank
Yards

Loose Yards = Bank
Yards + Shrinkage

Compacted
Fill

Figure 25.7

more fill material than the swell factor would indicate. It is better to have a quantity error slightly high rather than low, since it is usually more economical to dispose of a small excess than to purchase additional material.

In Figure 25.7, a pit to be excavated in sandy loam $25' \times 16' \times 12'$ would yield $178 \times 1.22 = 217$ cubic yards of loose dirt for disposal. Should this open pit be refilled with similar imported material, the quantity might be $178 \times 1.24 = 221$ loose yards. This small distinction between swell and shrinkage can be important in cases of very large quantities.

The superintendent, more than anyone, is responsible for the shoring of vertical earth trench and pit sides. Because it is time consuming and expensive, shoring is avoided when possible by sloping the sides. When it cannot be avoided, shoring must be designed so that it is not only adequate, but also allows for a large safety factor. Human safety is the first consideration; avoiding costly re-excavation is the second.

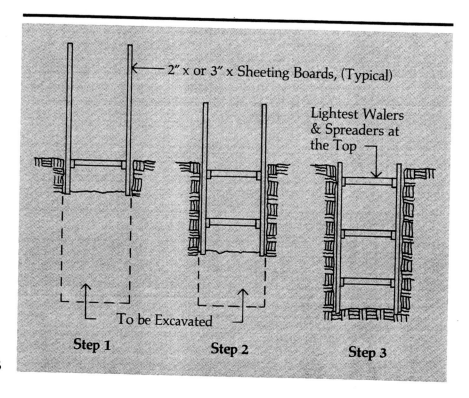

2" x or 3" x Sheeting Boards, (Typical)

Lightest Walers
& Spreaders at
the Top

To be Excavated

Step 1   **Step 2**   **Step 3**

Figure 25.8

Additional time and expense are required for the heavier shoring of unstable earth. The reason for this is excavation which must be done concurrently inside the shoring. Figure 25.8 shows how the first excavation is made to a depth of about two feet, then sheeting is placed and held by the top waler and spreader (step 1). In step 2, after further excavation, the sheeting is lowered and the second waler with spreader is placed. In step 3, the shoring is finished when the excavation is finished.

In some projects, movable shoring may be used to advantage. Movable shoring consists of a box-like structure which is lowered into the trench. When work is completed inside it, the box is pulled by machine power forward into the trench for another section of work.

In other projects, steel or special wood sheeting may be used with greater speed, economy and safety than any other method. Such shoring is often sublet to specialists for a firm, prearranged bid price, or on the basis of cost per linear foot unit.

Timbers for sheeting, walers, and braces may be cleaned, preserved and stored in the company permanent yard or warehouse for future use. It is good practice for the superintendent to provide the estimating department with cost records for future bidding. Following is an example of an acceptable format:

1. **Actual Cost of Shoring Unstable Earth Trench 6' deep, 2520' long**
   Contact surface trench sides 5,040 square feet
   5 reuses of material
   Material from yard, depreciation 2,000 bf @ .05 =   1,000
   New material, reuses and salvage allowed   300
   Hardware, nails, etc.   200
   Labor, from time records 7,200

   Total 8,700

   $$\frac{8,700}{5,040} = 1.73/sf., \text{ unit cost}$$

2. **Actual Cost of Shoring Stable Earth Trench 6' deep, 2100' long**
   Contact surface trench sides 4,200 sf
   Material from yard, depreciation 750
   New material (none) . . . . —
   Hardware 150
   Labor 3,000

   Total 3,900

   $$\frac{3,900}{4,200} = .93/sf, \text{ unit cost}$$

When underground water is encountered, there may be several approaches the superintendent can take. Some of the alternatives are:
- Ignoring the water—if it is not too deep
- Pumping out directly (A in Figure 25.9)
- Pumping from one or more sump pits (B in Figure 25.9)
- Providing drainage to a lower level (C in Figure 25.9)
- Installing a well point system (D in Figure 25.9)

The method which is the simplest, least expensive, and works well in the given situation is the best. The contractor should own one or more pumps, and the superintendent should keep standby pumps available in the event of breakdowns.

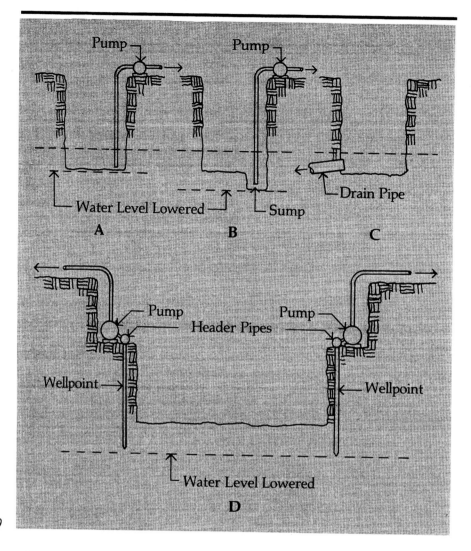

Figure 25.9

The wellpoint system (d in Figure 25.9) is expensive, but it is sometimes the only practical method for dewatering. The materials may be partly purchased and partly rented, or the entire operation may be sublet to a specialist. Perforated pipes are driven around the pit area and connected to horizontal header pipes which are attached to one or more pumps. This system is usually operated without shutdown for as long as the construction project requires. It is discontinued only after all waterproofing and backfilling of the structure is completed; the water table is then allowed to resume its natural level.

As in the case of shoring, the cost of dewatering is valuable feedback information for the estimating department. Unfortunately, it is almost impossible to convert dewatering costs to unit prices useful for estimating future projects.

The cost of dewatering depends on such variables as:
- Elevation of water table
- Rate of water entry
- Size of the excavated pit or trench
- Length of time the pumping must be continued
- Hours of operator time and pay rate

Another important factor when considering dewatering of any site is the effect that this process may have on the land and structures surrounding the construction site. There have been instances of adjacent structures settling and their foundations being damaged as a result of improper dewatering. Professional assistance is recommended in planning any dewatering project.

Chapter 26
Layout

The initial layout is usually, and wisely, assigned to a professional surveyor. Nevertheless, the superintendent will benefit from a close observation and understanding of the critical points, marks, and elevations.

The survey record should be kept intact and free of any additional markings. It should be stored in a locked, fireproof safe (see Chapter 6 for more discussion of the field office arrangement). Information that is taken from the survey record may be printed in colored ink on the project working drawings for quick reference.

Layout begins with the establishment of a reference plane, or bench mark. For construction purposes, the bench mark does not have to be based upon sea level (000.0 feet). It is usually selected for its relative permanence, independent of the ongoing construction activity. Typical bench marks are the finish floor of an existing structure, or a manhole cover in a street.

When the most critical surveying has been done, the superintendent may wish to take charge of the remaining detailed layout work, using conventional, company-owned or rented instruments and tools.

In *differential leveling*, the elevations of any and all desired points in a project are found by subtracting from or adding to the elevation of the bench mark as illustrated in Figure 26.1.

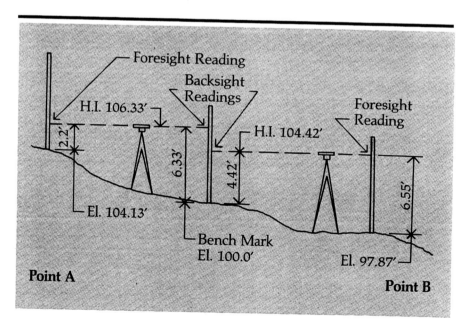

Figure 26.1

A *self-leveling* level saves a lot of time, since it eliminates the use of the tubular spirit level. When special accuracy is desired, the *theodolite*, which has an optical micrometer scale and optical plummet, may be used.

A *story pole* (SP) is often used for convenience, repetitive work and for marking up without defacing the higher quality and more expensive *leveling rod*. A typical story pole is a plain $1'' \times 2''$ board that can be marked as desired for a semi-permanent measuring stick. It can be used to good advantage in grading for concrete slabs, setting screed elevations and checking finish grades. Figure 26.2 shows a method for using the level and rod or story pole to establish a finish grade for an earth fill.

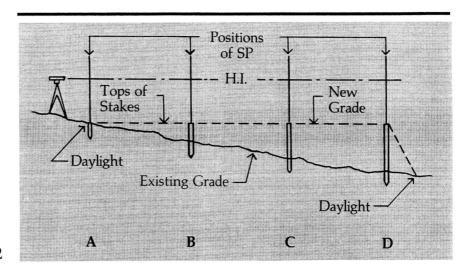

Figure 26.2

At A, a stake is driven flush with the earth surface, and the SP rests on it. The height of instrument (H.I.) is found and marked on the story pole. At other points, B, C, and D, stakes are driven until their tops, supporting the SP, bring the marked H.I. in line with the cross hair in the level. The tops of these stakes are sometimes painted blue and the stakes are called "blue tops".

Certain conventions for marking stakes are used in the surveying profession; for instance:

1. A crayon mark is made at ground level to ensure against accidental disturbance of the earth (A in Figure 26.3).
2. When the new grade, or subgrade (SG) is above the existing grade (EG) and fill (F) is required, SG and F are printed on the stake, with an arrow pointing down to a horizontal line at the desired elevation (B in Figure 26.3). Alternatively, the arrow may point upward to a mark at the top of the stake (C in Figure 26.3).
3. When the new subgrade is below the existing grade, SG, C (cut), and the depth of cut are printed on the stake with an arrow up or down to a horizontal mark (D and E in Figure 26.3).
4. Cut (C) or fill (F) information is written on one face of the stake; and location, such as station (9 + 00), centerline (℄), distance from centerline (50-R), toe of bank (TOE) and top of bank (TOP) is written on the other face of the stake (F, G, and H in Figure 26.3).

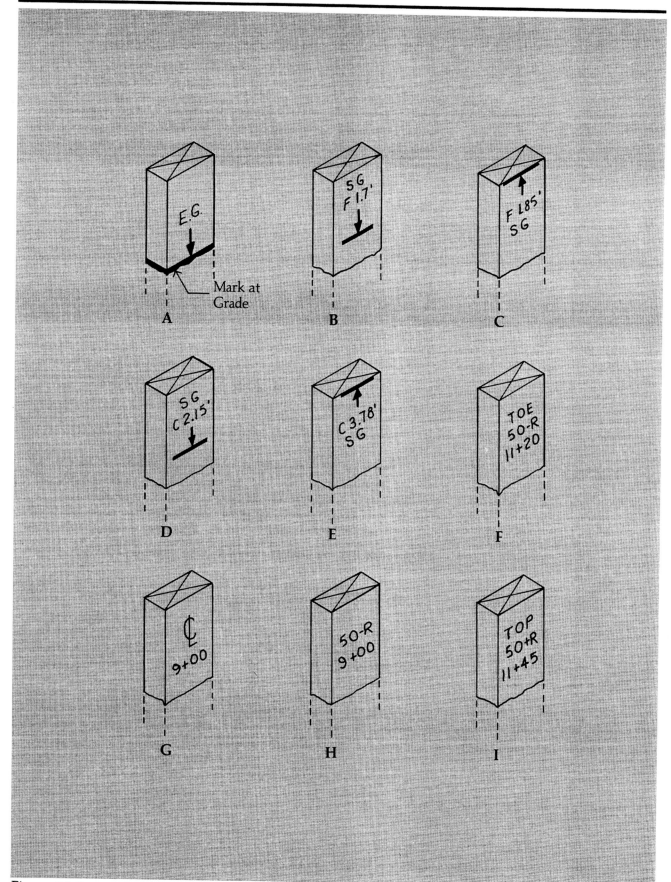

Figure 26.3

When the ground rises or drops steeply from the height of instrument (H.I.), *turning points* (TP) (as many as necessary) are taken, stair-stepping up or down the slope (Figure 26.4). TP's consist of one, or a series, of backsights (BS) and foresights (FS). The elevation of each TP is found by subtracting the FS from the preceding BS and adding the difference to the known elevation of the preceding TP. Once a series of TP's is made up or down a slope, a reverse check run is a good practice to ensure accuracy.

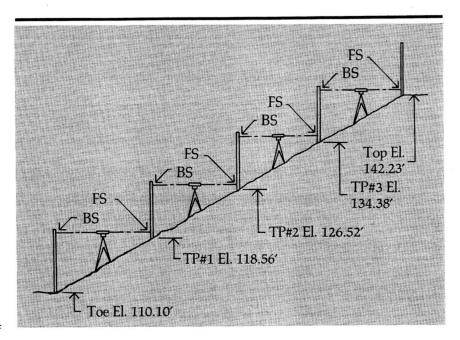

Figure 26.4

Knowledge of *bearings* and *azimuths* is valuable to the superintendent for locating property line distances and corners. In Figure 26.5, the property stake at corner D is missing. Given the bearings on a plot plan, the superintendent, with an assistant, sets a transit over C, orients to South, and sights along the bearing S 48 degrees —45′W (see Appendix—"Surveying Techniques"). He stretches a string between stakes at C and some arbitrary point X. Then he sets the transit over corner E and sights along bearing N 43 degrees —20′W and stretches a string from E to some arbitrary point Y.

D is assumed to be the intersection of CS and EY. Secondary proof is a check measurement of all five property lines; final proof is a check for closure as follows:

1. Compute the azimuth of each bearing (see Appendix for rules).
2. Determine all interior angles (see Appendix). 3. The sum of all interior angles should equal the total number of degrees of those angles. This refers to a closed figure having a given number of sides, using the formula N − 2 × 180 degrees (where N = the number of sides). In Figure 26.5, 5 − 2 × 180 degrees = 540 degrees and

$$
\begin{array}{r}
165° — 30' \\
120° — 00' \\
76° — 35' \\
87° — 55' \text{ Check} \\
\underline{90° — 00'} \\
540° — 00'
\end{array}
$$

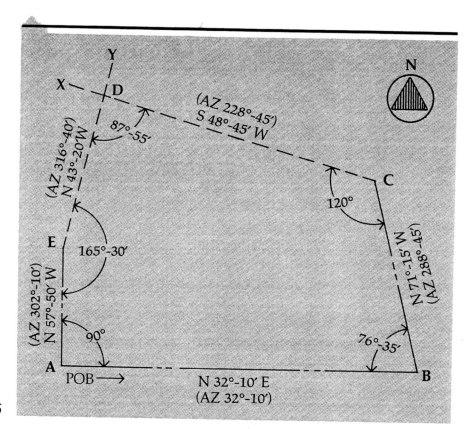

Figure 26.5

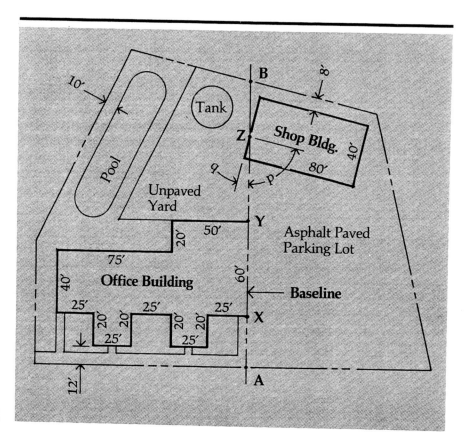

Figure 26.6

146

The layout for one or more structures on a site is more conveniently carried out with a baseline for reference. If a baseline is given in the drawings, it is advisable to use it; otherwise, the superintendent and his layout crew can choose a suitable line, as in Figure 26.6

Points A and B are semi-permanently marked on property lines. Points X, Y and Z are staked on the baseline for bench marks to control all other layout for the office building and shop building. All lines for the office building are parallel or perpendicular to the baseline; lines for the shop building are p degrees and q degrees to the baseline. Distances AX and XY are found by simple addition of the given dimensions. Distance BZ is found by trigonometry, and ZY is ignored.

In the event that stakes are lost, *reference marks* simplify the restoration work. In Figure 26.7, after the corners of a structure are staked, temporary marks are made at distances and locations where they are relatively safe from disturbance. If, for instance, corner stake Y is lost, a level of transit can be set at either point A or X and sighted on B; a pole can be moved into the telescope line of sight, and at a taped distance of 50' from X. Thus Y can be located and restored.

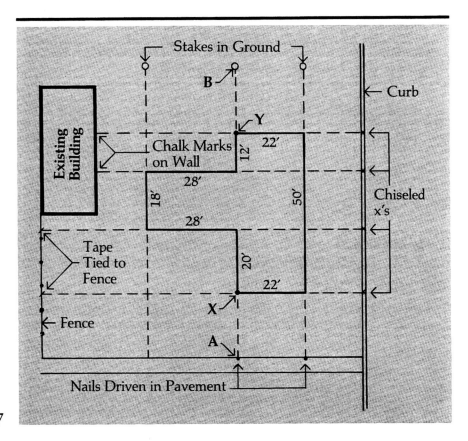

Figure 26.7

Another way to preserve corner locations and elevations is with the *batter board* system. Since excavating and forming for footings will displace corner and intersection stakes, offset stakes (a very ancient technique) are the answer. Figure 26.8 shows a typical arrangement. If all batter boards and strings are placed before the excavation work, machine work would clearly be impossible, and handwork greatly hampered. To avoid that complication, only the main corners and intersections are batter boarded, the strings removed, and the excavations done. The remaining batter boards and all strings are then placed. Trenches can be hand shaped as required. The horizontal batter boards are set at selected elevations above trench bottoms, or footing tops, so that continual vertical check measurements may be taken. Since strings lose their tension with temperature changes and physical disturbances, a simple and easy way for regular tightening is the use of C-clamps, as shown in Figure 26.8.

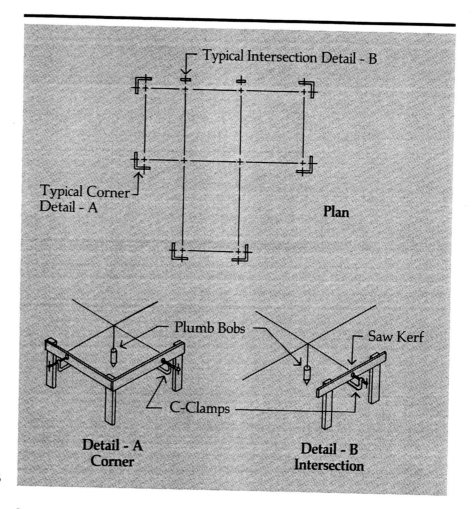

Figure 26.8

It is not practical to batter board some points, such as those at interior footings C and D, at this time. These can usually be postponed until a certain amount of footing work is in place, and then they can be located by special measurement, marking and/or staking.

On any project, leveling instruments can be in short supply and alternative ways of measuring must be used to avoid lags in the progress schedule. Following is a list of some other methods; they appear roughly in order of accuracy, beginning with the most reliable:

1. *Straight board and carpenter's level* are limited to short distances of 20′ or less.

2. *The string level*, Figure 26.9, is only as accurate as the tension of the string permits. It is most accurate when positioned at the center of the string.

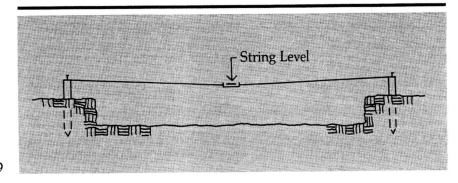

Figure 26.9

Some improvement in accuracy can be obtained with intermediate supports, as shown in Figure 26.10

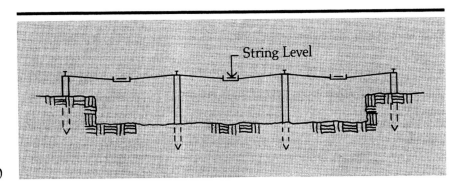

Figure 26.10

3. *A water hose level*, as in Figure 26.11 can be useful in unusual conditions such as when visibility is blocked.

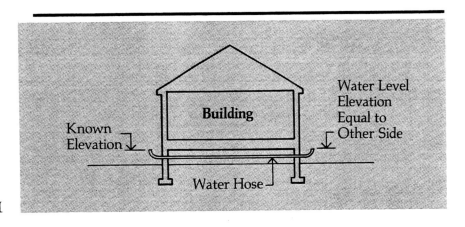

Figure 26.11

4. Direct sighting, as in Figure 26.12, might be acceptable in some cases where accuracy is not important. Height h1 is measured from the ground to the line-of-sight; h2 is then measured from the ground to the mark on the tree.

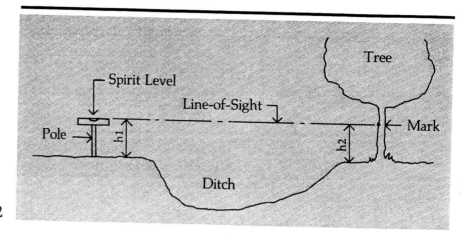

Figure 26.12

5. The hand level produces a rough approximation of elevations. Eye level is equivalent to height of instrument (H.I.) and varies with the individual user. In Figure 26.13, H.I. is 5'-2". Working up or down a slope with the hand level is similar to the procedure with the engineer's level, Figure 26.4, the main difference being the degree of accuracy.

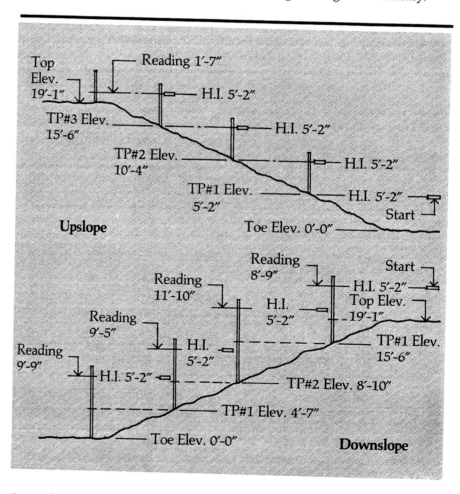

Figure 26.13

In circular or semicircular structures, a semipermanent swinging arm can effectively be used for the layout, as shown in Figure 26.14

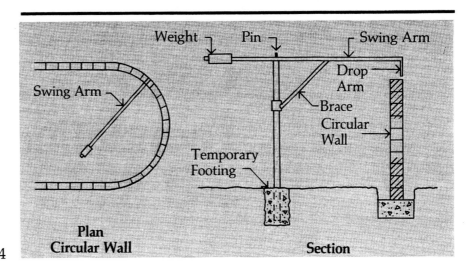

Figure 26.14

Plan Circular Wall   **Section**

The curvature layout can be maintained by this single temporary instrument for the excavation, footing, formwork, foundation wall formwork, block masonry work, and concrete bond beam formwork.

Chapter 27

Concrete Footings, Foundations, and Slabs

Before starting any concrete work, a supplier must be determined and approval obtained for each mix design to be used on the job. The concrete supplier will want to know the approximate yardage for each concrete mix and will require a tentative schedule. The business of ordering and figuring quantities is discussed in more detail later in this section.

Footing Work

Footing work begins with the layout. Chapter 26 briefly covers the subject of batter boards, used to temporarily locate corners, intersections, and lines to guide the excavation work. For speed and economy, the greater the proportion of machine work to hand work, the better. It is also desirable to minimize the quantity of excavation and backfill. If the earth lacks the stability for vertical cuts, it may be sloped and the footing sides formed. If the earth is stable, less excavation, less backfill, and no side forms are required, but more hand trimming is involved. A similar choice exists with integral footings and slabs.

For very deep footings, excavation and backfill can be enormous and costly. The work and expense can be reduced slightly by benching; but even if the earth is stable enough for vertical cutting, greater width may be necessary to provide room for installing the formwork. OSHA safety requirements call for shoring all trench excavation with vertical sides over 5' deep (see Chapters 24 and 25 for more information on clearing, earthmoving, and mass excavation). All of these methods assume that footing excavations are made down from the final (finish) grade. Filling might also be done after the foundation walls are in place. This method is worth considering when there are many closely spaced foundation walls, piers, and footings.

Structural excavation can be minimized by filling inside the building walls first, and sloping the sides of the fill. Structural excavation is made for footings only. Filling around the building is done later by machinery.

Footing side form designs differ depending on whether the footing tops are above or below grade and the construction joints for the foundation walls are flush with the top or raised. The accuracy of footing lines, levels, and elevations begins with the batter boards and stretched string or wires; it is further checked and controlled by engineer's level and rod. The diagonal measurements should also be checked.

In a small project, the superintendent may order enough forming materials to construct all footings at one time. In a very large project, the footing materials are calculated, and then a fraction of that total amount is purchased and reused repeatedly. Materials for footings and foundation walls are usually calculated and purchased together, since their construction overlaps. An example calculation of quantities follows the discussion of foundation formwork.

Foundation Formwork

Foundation formwork requires two considerations not usually given to footings: (1) engineering for stress, and (2) effect on the surface appearance of the concrete (if it will be visible). The main factor in stress design is the height of the wall. Foundation walls or "stem walls", typically include continuous footings. Grade beams are foundation members supported only at their ends.

In Figure 27.1, (A) is an example of formwork high enough for the use of walers, (B) is foundation formwork low enough not to need walers. The higher the wall (C), the heavier and closer together the formwork members must be in order to resist the increasing pressure of the wet concrete.

The principles of formwork design are too varied and complex for the scope of this book, but some of the important points are:
- Aim for safety, quality and economy.
- Specify the type and quality of concrete surface desired.
- Include in your design considerations:
  a. Rate of concrete placement (ft. per hr.).
  b. Temperature at time of placement.
  c. Height of wall.
  d. Number of material reuses. Many reuses warrant better quality materials, workmanship and the prefabrication of panels to avoid damage and waste.

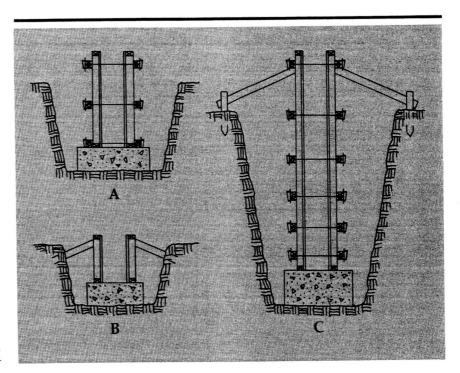

Figure 27.1

Tables of standards are available to the superintendent to guide the selection of form members, their spacings and the sizes and number of ties. Occasionally, more complex formwork may require the services of a professional engineer.

Figure 27.2 shows a system for calculating the quantity of forming materials for purchase based on the greatest number of reuses possible without slowing job progress. From the given date, lumber and plywood required for the first concrete pour would be: 40/320 = 1/8th of the total formwork, or 840/8 = 105 linear feet = 4200/8 = 525 board feet. However, formwork cannot be stripped and reused for three curing days; therefore, lumber to form three consecutive days worth of concrete delivery would be: 525 × 3 = 1575 board feet.

There would be five incremental moves with some waste and breakage occurring each time, for which an allowance of, say, 10% per move should be added. Thus:

Initial quantity	1,575 bf
Waste 525 × 5 × .10 =	263
	1,838 bf

This is an optimistic schedule, and the superintendent would be wise to purchase material for an extra increment, for a total of: 1,838 + (525 × 1.10) = 2,416 bf. Leftover material may be used in other concrete formwork in the project.

When footing and foundation walls are placed separately, the system outlined above may be used. The footing sequences should be scheduled to lead the walls by one or more days.

Keeping within the cost budget is a subject discussed in Chapter 13— "Cost Controls". In Figure 27.2, eight days are scheduled for erecting the formwork. If five more days are allowed for prefabrication and final cleanup, the total is 13 working days. Given the cost budget of $20,160 and the average pay rate of $19, the number of workers would be:

$$\frac{20,160 \text{ (\$ in cost budget)}}{19 \text{ (\$/hr.)} \times 8 \text{ (hrs./day)} \times 13 \text{ (days)}} = 10.2 \text{ (workers)/2 (days total)}$$

The superintendent tries to strike an overall average of 10 workers. If he can keep within the 10 worker average for 13 days, he can be assured of successful cost control.

Before placing the concrete, other preparations include:

- **Confirming concrete mix design approvals.** The project specs provide minimum requirements for the concrete, including:
  - Fine and coarse aggregate sizes and characteristics
  - Type of cement (I, II, etc.)
  - Strengths in pounds per square inch (psi) for use in various locations
  - Admixtures
- **Applying bond-breaking oil.** Form oil should be applied to surfaces which will contact concrete, to aid in stripping and to preserve the form materials through several reuses.
- **Forming out for openings and bulkheads.** Openings must be formed for such items as ventilators, access doors and windows, and for bulkheads (Figure 27.3) before formwork is buttoned up.
- **Setting inserts.** Inserts such as anchor bolts and pipe sleeves should be carefully positioned in the forms. Subcontractors should be given the opportunity to provide and install inserts for their own trades.

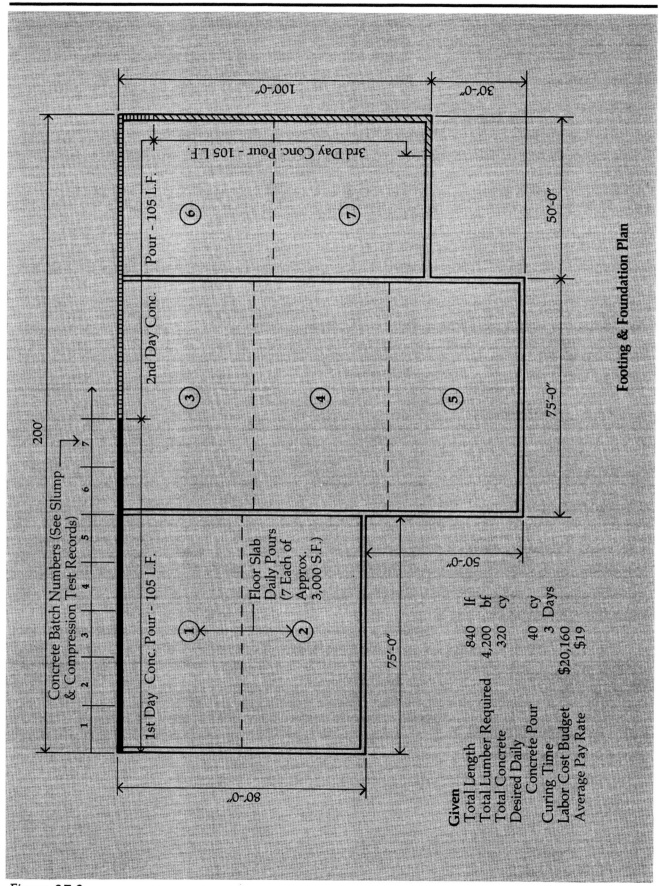

Figure 27.2

Footing & Foundation Plan

Given

Total Length	840	lf
Total Lumber Required	4,200	bf
Total Concrete	320	cy
Desired Daily Concrete Pour	40	cy
Curing Time	3	Days
Labor Cost Budget	$20,160	
Average Pay Rate	$19	

155

- **Obtaining inspections.** An inspection of forms by the sponsor and various government agencies should be arranged and their approvals recorded.
- **Wetting down the earth.** Earth surfaces against which concrete is to be placed should first be soaked to prevent excessive absorption of moisture from the concrete. In hot weather, the formwork should also be water-soaked.

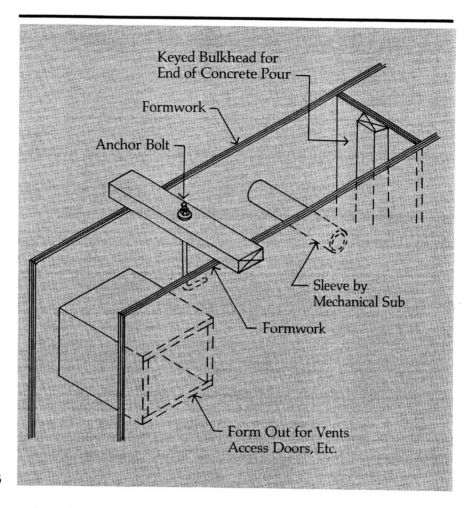

Figure 27.3

In the early stages of the project, the superintendent should secure a source of concrete and obtain approval from the sponsor of the mix designs. He should give the concrete supplier a rough schedule of quantities and dates for deliveries.

If the exact calculated volume of concrete were ordered, there almost surely would be a shortage resulting from one or more of the following:
- Earth under footings or slabs may not be graded to true elevations.
- Footing side forms may be omitted and concrete poured against the earth trench sides, which are irregular.
- Forms that leak or are out of line (bowed).

- The necessity to replace hardened concrete (due to delayed pours) with fresh concrete.

Footings with side forms	80 cy + 11% =	89
Footings without side forms 14″ × 10″ with 3″ overcut	60 cy + 27% =	76
Foundation walls	180 cy + 11% =	200
Total quantity to order		365 cy

An order to the transit mix supplier for concrete could appear as follows:

Description	Delivery, Beginning at 8:00 a.m.	
Strength 2,500 psi, 28 days	April 6, 1986	46 cy
1″ maximum aggregate size	April 7, 1986	46 cy
Type I cement	April 8, 1986	46 cy
Air entrainment	April 9, 1986	46 cy
6 gal./sack water max	April 10, 1986	46 cy
6.3 sacks cement/cy min	April 11, 1986	46 cy
	April 12, 1986	46 cy
	April 13, 1986	43 cy
	TOTAL	365 cy

A number of factors affect concrete strength, but the main factor is the amount of water (gallons per sack of cement). The strongest possible mixture would be too dry to use. For more plasticity (workability), some strength must be sacrificed and vice versa. It is important, therefore, for all concrete deliveries to be as consistent as possible. One conventional method of maintaining consistency is the slump test, using the slump cone, Figure 27.4.

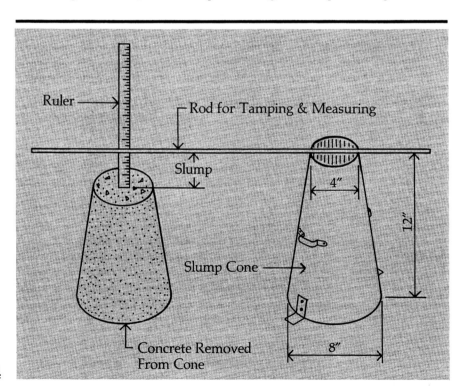

Ruler

Rod for Tamping & Measuring

Slump

4″

12″

Slump Cone

8″

Concrete Removed From Cone

Figure 27.4

At least one test should be made from each batch (truck load). The concrete is placed in the cone in three layers, and each layer rodded with 25 strokes, each stroke penetrating the previous layer. The slump that corresponds with the ideal plasticity is used as the standard for all other batches. In subsequent tests, lesser slumps indicate that a batch is too dry, and greater slumps indicate that it is too wet. When a deviation from the desired standard slump occurs, it is usually too late to correct the batch, but the following batch may be adjusted. It is good practice to record the location of each batch within the structure, along with their respective slump test results. (see Figure 27.2).

Even with the most faithful adherence to the design mix and to good rules of placing, vibrating, and curing, there is no absolute assurance that the specified strength will be obtained; therefore, architects and engineers often demand follow-up testing. A conventional test is one in which samples are taken from delivered batches of fresh concrete, stored in standard size cylinders and crushed by laboratory equipment at seven day and 28 day intervals. The strength in psi is thus determined. The samples are taken from the various batches which were used. Once again, the batch locations are recorded (Figure 27.2). Any concrete batch which is found (by the sample) to be under-strength, can be identified and steps taken to reinforce or replace it.

Concrete can be tested after hardening by means of core samples, but this is an expensive method which the superintendent tries to avoid. This procedure involves cutting cylinders from the concrete with special coring equipment and then crushing these core samples to determine their strength under compression.

It is important for the superintendent to attain the specified minimum strengths of concrete in order to avoid structural collapses, whether or not legal liability exists.

Although the superintendent may obtain nearly all of the project's concrete from transit mix sources, there are occasions when he needs small quantities mixed on the job. When precise specifications are not required, a convenient formula is the 1:2:3 ratio (1 part cement, 2 parts sand, 3 parts gravel) and six gallons of water per sack. The total loose quantity is 1-1/2 times the net, in-place quantity.

Example: 20 cy are required $\times$ 1.5 = 30 cy loose material

Cement	$30 \times 1/6 =$	5 cy = 135 sacks
Sand	$30 \times 2/6 =$	10 cy
Gravel	$30 \times 3/6 =$	15 cy
Water	$135 \times 6 =$	810 gallons

Concrete placed in footing and foundation forms is most often and most economically done directly from the chutes of transit mix trucks. Locations inaccessible to trucks may require wheelbarrows or buggies, and possibly temporary runways and bridges. For large quantities, it may be economical to (1) make temporary ramps and bridges for truck access, (2) use concrete pumping equipment, or (3) use a crane and bucket. The final choice results from a careful estimate of time, costs, and equipment availability. Following is an example (omitting details of the computations):

Given 300 cubic yards to be placed,

Method 1—Using wheelbarrows or buggies and runways:

30 Hrs., $2.50/cy = 750.00.

Method 2—Using earthmoving equipment and handwork, making and later removing ramps and bridges:

12 Hrs., $2.20/cy = $660.00.

Method 3—Using rented pumping equipment:

8 Hrs., $2.00/cy = $600.00.

In this example, pumping proves to be the most economical in terms of both time and cost. However, the same might not be true in every case.

Reinforcing steel is a very important part of concrete placement. The superintendent has the following responsibilities regarding reinforcing steel:

1. Coordinating the formwork construction with the placing requirements of the the rebar subcontractor.
2. Arranging for concrete placement in form spaces that are crowded with rebar.
3. Ascertaining the proper spacing and location of rebar in the form.

To avoid segregation, the concrete should be lowered into the forms as gently as possible and in a manner requiring a minimum of spreading by shovel or vibrator. Concrete must be consolidated in the forms in order to eliminate voids, fill all corners, and to tightly bond it to the rebar. The concrete should be placed in layers, or lifts, not more than 12 inches thick, and each lift consolidated by one or a combination of the following methods:

- Hand spading tool
- Internal mechanical vibrator
- External mechanical vibrator

In many cases, there is not enough form surface for effective use of the external vibrator. It is good practice to combine hand spading and internal vibrating. As a rule, vibrating should be done at intervals of 16 to 20", and for a period of from 5 to 20 seconds at each spot.

Curing of footings and foundations is most easily done by keeping the forms in place and continutally wetting down with water. When, as in Figure 27.2, the forms must be removed before the typical seven day curing period is completed, curing may continue with wetting, covering with burlap or plastic membrane, or spraying with a water-retaining chemical coating.

Pointing and patching of surfaces which will be concealed from view (by earth backfill, for example) might not be required at all, except to ensure rebar coverage. Visible surfaces (above grade) may be specified to receive anything from simple patching of honeycombs to an architectural rubbed finish, as discussed in Chapter 28, "Superstructure Concrete Work".

Slabs on Ground

Slabs on ground can be divided into the following typical operations:
- Performing the layout
- Fine grading the earth subgrade
- Laying aggregate base courses, if any
- Constructing edge (header) forms
- Laying waterproof membrane, if any
- Placing inserts, if any
- Placing interior construction joints and depression forms
- Setting screeds
- Installing rebar and/or mesh (by subcontractor)
- Furnishing and placing concrete
- Finishing the concrete surface
- Special finishing and jointing, if any
- Curing and protecting

Layout for floor slabs (sidewalks and other slabs are discussed in Chapter 29, "Precast Concrete Work") is usually straightforward, since footings and foundations already exist to define the perimeters and approximate elevations. It can become complicated when there are numerous depressions and slopes to floor drains. These complications are coordinated with the work of setting screeds, which is discussed later in this section.

Before anything else, layout is necessary for the fine grading of the earth subbase, which is usually left irregular by the earthwork subcontractor and by foot and light equipment traffic. Figure 27.5 shows an example of the layout required for a floor slab.

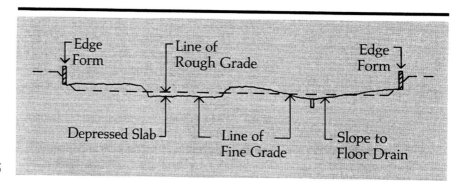

Figure 27.5

After layout, fine grading (consisting of shallow cuts and fills) can be done by hand or machinery, or a combination of the two. Tamping by pneumatic equipment is advisable in fill areas to avoid settling under the weight of the concrete and live loads. Perfectionism is most important when the slab goes directly on the subgrade; it is less crucial when there is a course of aggregate subbase. The surface of the aggregate then receives the most careful fine grading. Since concrete is expensive, it is important that thickness requirements are carefully met. When the slab has shallow thickened edges, excavation is a part of the fine grading work; deeper excavations are a part of the structural (footing) excavation.

Edge forms may then be placed to true lines, elevations, and levels. One or more layers of aggregate may be compacted thoroughly, preferably by vibratory compactors which shake the particles together tightly. The surface of the aggregate should be as accurate to plane as possible in order to conserve concrete material (see Figure 27.6). One or two inches of sand placed over the waterproof membrane helps prevent puncturing or tearing during the placing of mesh, accessories, and concrete.

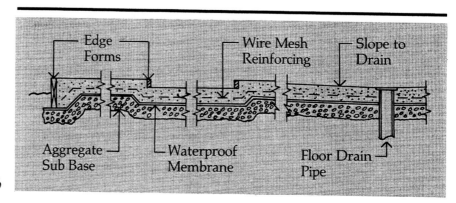

Figure 27.6

Interior forms, such as those for construction joints and depressed slabs, are coordinated with the setting of screeds. Whenever the superintendent has a choice of location, he will place construction joints to coincide with the end of a day's concrete pour. Figure 27.7 shows typical floor joints. A keyed construction joint is shown at (a). When the form is removed, the next pour is "keyed" to the old, preventing a line of settlement. A control or weakened plane joint is shown at (b), which permits a controlled crack to open and close with contraction and expansion. An isolation or expansion joint is shown at (c), with compressible filler material and poured sealant. These joints can occur at walls, columns, and equipment bases which are rigid relative to the slab. When the joints occur between two slabs (d), a form is used to hold the filler material. A removable strip makes possible a depression for the sealant (e) after the form is removed and the second slab placed.

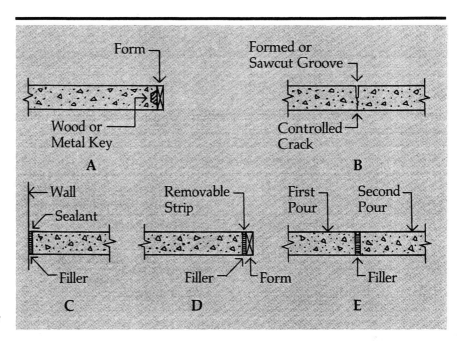

Figure 27.7

The setting of screeds (Figure 27.8) is the customary method of controlling the slab thickness and the accuracy of the top surface elevations. Because of the natural settling and shrinkage of concrete as it cures, it is a good idea to set the screeds slightly higher (camber) than the desired finish elevation. As the concrete is placed, crews drag the screed board along the 2" × 4" guide strip in a see-saw motion to achieve the desired plane. On very large slabs, a mechanical vibratory screed may be used to allow a lower slump of concrete and to increase the density. On some projects, wet screeds are allowed; this method can greatly increase the overall operation of placing screeds and concrete.

Transit mix concrete may be scheduled for delivery in a manner similar to that previously described for footings and foundations. The quantity to be poured each day depends on the progress schedule, previously designed. The optimum size of the crew, particularly the finishing crew, must be determined based on the desired amount of daily placement. The method of conveying the concrete from the transit trucks must also be determined in order to keep up with the desired daily placement goal.

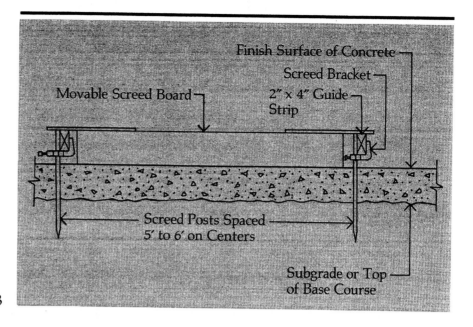

Figure 27.8

For the following example, refer to Figure 27.2, which is divided into seven sections. The dotted lines represent construction joint header forms. Given seven days for completion, what would be the optimum size finishing crew, assuming an average production of 560 square feet per finisher per day?

$$\frac{20{,}750 \text{ sf total area to be finished}}{7 \text{ days} \times 560 \text{ sf—per finisher per day}} = 5.29, \text{ or, say 5 finishers}$$

What quantity of concrete should be ordered and placed daily for a 4" thick slab?

Net quantity $\dfrac{256 \text{ cy total}}{7 \text{ days work}} = 37$ cy per day

Add for waste $\underline{4 \text{ cy}}$

Total 41 cy per day

What is the best method of conveying the concrete? A comparison of costs for different placing methods finds the following for slabs on ground:

Method 1—Pumping . . . . $18.60/cy $765
Method 2—Crane and bucket . . . . $21.00/cy $860
Method 3—Barrows/buggies and runways . . . . $19.30/cy $790

Pumping appears to be the most economical in terms of both time and cost.

The total placing and finishing crew might be:

Pump hose tenders	2 @ 4 Hrs.	= 8 Hrs.
Shovelers	1 @ 4 Hrs.	= 4 Hrs.
Screed operators	2 @ 4 Hrs.	= 8 Hrs.
Finishers	5 @ 6 Hrs.	= 30 Hrs.
	Total	50 man hours

$$\frac{50 \text{ man-hours} \times \$20 \text{ (avg. pay scale)}}{41 \text{ cy (total cy used in one day)}} = \$24.39/\text{cy}$$

Special finishes are specified in some projects. Some of the choices include color, exposed aggregates, scored patterns, stamped patterns, burnishing, and rock salt. The superintendent might have difficulty hiring competent finishers for such work and may consider subletting.

Curing should be maintained for at least seven days by the most suitable method. The alternatives include:

- Keeping wet with water
- Spraying with a chemical sealer
- Covering with earth, sand, or straw
- Covering with paper or plastic
- Covering with burlap that is kept wet

Perhaps the most commonly used curing method for slabs is a sprayed-on water-retaining membrane which eventually disintegrates. If there will be traffic on the slab while it is curing, heavy paper and/or sheets of plywood are suggested for protection.

Chapter 28

Superstructure Concrete Work

The superstructure is the portion of a building that extends above the foundation. This section is directed to the conventional components found in almost every concrete building, namely:

- Walls above grade
- Columns
- Beams and girders
- Spandrel beams, parapets, and bond beams
- Suspended slabs
- Suspended stairs

Concrete Walls Above Grade

Concrete walls above grade are constructed in the same way as the high foundation walls described in Chapter 27. Those that are high enough to be exposed to view require heavier forming lumber, and may also require form liners for architectural effects. For large quantities, it is economical to construct panels that are uniform in width and height; this approach minimizes the need to redrill for ties. It is common practice for one side to be erected with primary bracing. There is then a delay while the subcontractor is placing reinforcing steel; and finally, the other side forming is erected with secondary bracing.

Figure 28.1 shows typical details; (a) is a snap tie which also acts as a spreader; (b) is a plastic or wood cone coil tie; (c) is a method of connecting wall form panels together; (d) is a method for supporting a scaffold on the form side for use by workers; (e) is a method of holding a single side form against earth, rock or other material; and (f) is a method of holding form corners together against the pressure of the wet concrete.

It is worth noting that the superintendent has a number of choices other than the conventional forming method discussed here; the alternatives include:

- Prefabricated forms
- Gang forms
- Slip forms
- Flying forms

The superintendent should evaluate each of these systems when planning the formwork for his project. On the days when forms are stripped, moved and re-erected, the superintendent should reserve the use of small machinery, such as fork lifts, backhoes, bucket loaders, or cranes to handle the heavier panels.

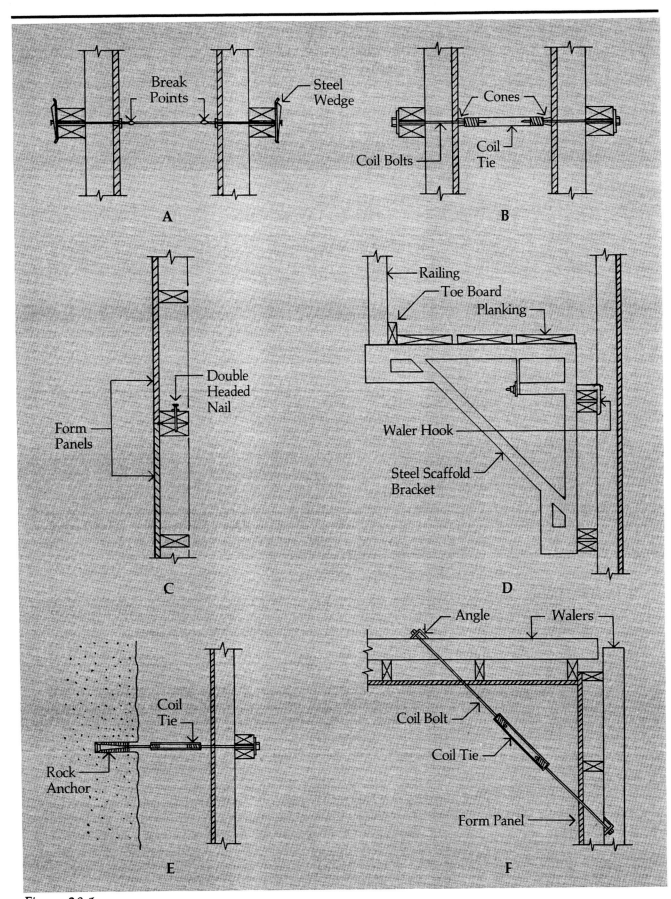

Figure 28.1

The materials to be purchased for footings and foundations may be calculated as outlined in Chapter 27. The number of reuses depends on the quality of the material being used for the formwork. The amount of formwork material needed depends on: (1) the number of anticipated uses from the material chosen, (2) the total quantity of formwork, (3) the number of days the forms must remain in place (curing time), (4) added allowance for waste, (5) deductive allowance for salvage of used materials existing on the site, (6) the time allotted in the progress schedule, and (7) the finished surface of the concrete required by the specifications. The following summary shows an example of calculations for material purchase.

Total quantity (if all were formed at once) .... 8,000 sf
Number of curing days .... 7 ea.
Total days allotted in progress schedule .... 30 ea.
Try increments of.. 1,000 sf
Total number of increments................................. 8 ea.
Number of reuses.. 5 ea.
Board feet of lumber per square foot of form................. 3 bf

Purchase three increments   3,000 sf × 3.0 bf per sf = 9,000 bf
Add waste   9,000 bf × 5 uses × .10 waste = 4,500 bf
Deduct for material on hand, say   (2,500) bf
                                      Purchase   11,000 bf

Total Time

Designing and fabricating	4 days
Erecting and reusing	24 days
Final stripping and cleanup	2 days
Total	30 days

This arrangement just barely complies with the progress schedule; the work might be done in less time at a higher cost. Cost control may be conducted in the manner previously discussed. If the labor budget is $24,000 and the average pay scale for the crew is $20 per hour, then:

$$\frac{\$24,000 \text{ (labor budget)}}{30 \text{ (working days)} \times 8 \text{ (Hrs. per day)} \times 20 \text{ (\$ per Hr)}} = 5 \text{ workers}$$

If the superintendent can accomplish the wall formwork within 30 working days with no more than five workers on average, he can keep the actual cost within the budget. The placing and finishing of the concrete is discussed later in this section.

Columns

Columns may be found in a variety of shapes, but this section deals only with the basic three types: square, rectangular, and round. A column is of such structural importance that the superintendent must make sure the vertical axis is correctly located, plumb, and securely held in place against movement. The subcontractor usually stands the rebar in cages and then the forms are built around, or lowered over the rebar. The bracing must be rigid and preferably four-way. Figure 28.2 shows three kinds of column forming: (a) round columns using fibre forms, (b) square or rectangular using wood and bolt yokes, and (c) square or rectangular using adjustable clamps. In the latter case, the clamps are arranged so that alternate corners are connected and tightened in order to avoid twisting the forms out of square. It is good practice to make the forms one half inch or so higher than the desired height dimension to allow for shrinkage of the concrete as it cures.

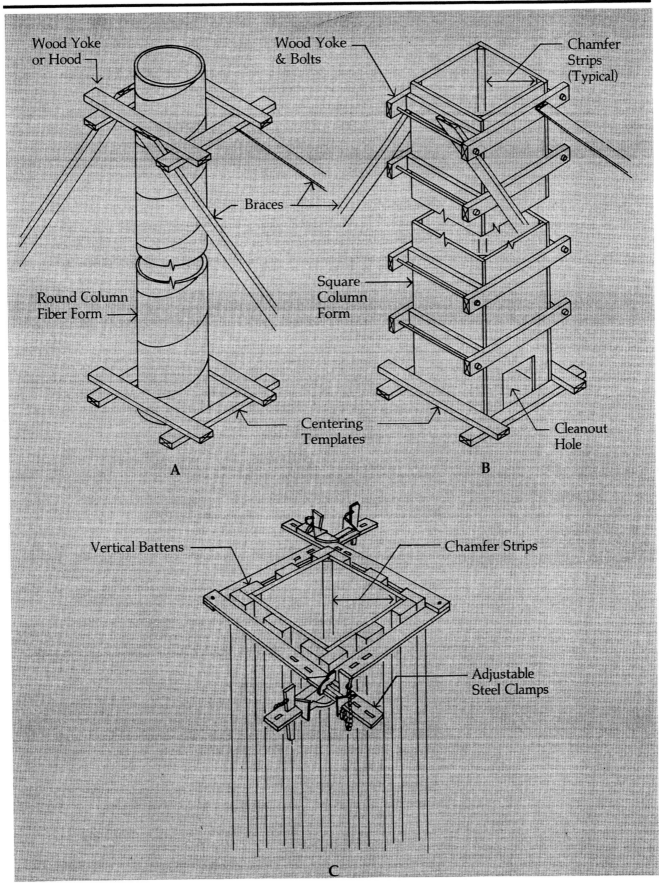

Figure 28.2

Chamfer strips placed in the corners produce concrete that is relatively free of irregularities, and therefore requires little, if any patching. Openings are usually made at the base of column forms for cleaning out debris. These openings are then closed before concrete is placed. For tall columns, openings are sometimes made at one or more levels in the forms. This is done in order to provide vibrators to consolidate the concrete as it rises.

Round columns may be made of wood or steel; but in recent years, factory made fibre forms have become the most popular choice. Unlike job-made forms which can be designed to hold against any desired rate of concrete pour, fibre forms, being stock sizes, put a certain restriction on the rate of pour. For instance, one brand specifies the average pressure causing rupture in an 18″ diameter tube at 447.9 pounds per square inch. Although it is unlikely that rupturing or bursting would occur, it is a good idea for the superintendent to be aware of the limitations of the forms.

Fibre forms cannot be reused, but job-built wood forms for square or rectangular columns may be reused for economy—if there are several columns of the same size. For instance, if there are 16 columns 12″ × 12″, and 16 columns 16″ × 16″, they might be formed with four of one size and four of the other, each form being used four times. The concrete for the column construction could be scheduled for placing at the same time as the wall sections.

Cost control can be computed as before: the budget divided by working hours times the average pay scale equals the number of workers required. If the number falls short of an adequate crew, it is possible in some cases for the superintendent to hire more workers, thereby completing the work in fewer total hours. Thus, given a budget of $8,000, a time schedule of 16 days and an average pay scale of $20,

$$\frac{\$8,000 \text{ (budget)}}{16 \text{ (days)} \times 8 \text{ (hours per day)} \times \$20 \text{ (avg. pay scale)}} = 3.125 \text{ workers}$$

In this case, the superintendent attempts to complete all formwork with three workers within 16 days, or four workers in 12 days.

Beams and Girders

Beams and Girders are one of the highest unit costs of all forming, because of the following factors: (1) the intricacy of the layout, (2) complexity of intersecting and end joints, (3) support structures, such as shoring, (4) limited reusability of materials, and (5) the limited scope for prefabricating and panelizing. It is important, therefore, to plan and detail beam and girder construction carefully in order to keep wasted material and motion to a minimum. Beam and girder forms are often made integrally with suspended slabs.

The shoring materials offer maximum reuses, but the beam and girder side and bottom forms for small quantities and varied sizes may have only one use (for large and uniform sizes, there might be a small number of reuses). Some economy is possible through the use of materials left over from footings, foundations, walls, etc. In the end, a lot of the beam form materials become useless scrap.

The labor cost control may be as follows: Given a budget of $20,400, a time schedule of 21 days, and an average pay scale of $20 per hour,

$$\frac{\$20,400 \text{ (budget)}}{21 \text{ (days)} \times 8 \text{ (hrs./day)} \times \$20 \text{ (avg. pay scale)}} = 6.07 \text{ workers.}$$

In this case, the superintendent may try to accomplish all beam and girder formwork with six workers within 21 working days.

Because of the required curing time, the bottom form and shoring might remain in place 14 days or longer, while side forms may be stripped for early reuse.

It is quite common in beam form construction to camber them slightly (raise the elevation at the center one half inch or so) to compensate for the sagging under their own weight, a common condition after concrete is poured and the shoring removed.

Spandrel Beams, Parapets, and Bond Beams

Spandrel beams usually occur above or below continuous wall openings, such as for windows. They may or may not be of great structural importance. Spandrel beams usually have architectural significance and often receive special finishes.

Parapets are curbs around the perimeter of roof slabs. They serve as safety railings and provide architectural effects. Bond beams are structural, and sometimes architectural, concrete features in masonry walls. Bond beams are sometimes used as fillers between masonry wall tops and sloping roofs. Forming of these special beams follows the principles of regular beams and girders, but bond beams are more likely to be cased integrally with suspended slabs. Bond beams present a special case of forming, since they are usually supported on masonry walls and have no bottom forms.

Suspended Slabs

Suspended slabs for floors and roofs can be the least expensive type of formwork in terms of unit cost, particularly when they cover large, uniform expanses in simple detail. Since joists and sheathing are common to most slab forming, economy lies in the best choice of shoring method. A traditional method of shoring is with vertical, braced wood or metal shores. The choice of wood or metal depends on a number of conditions, such as the height of the shoring, ownership or rental by the contractor, and the speed of turnover in reuses. Metal shores have the following advantages over wood: (1) long life, (2) adjustability, (3) lower labor cost to erect and remove, and (4) if not greater strength, at least predictability of inner resistance to stress.

As many forming members as possible can be removed from below after the concrete has cured sufficiently. However, some individual shores might have to be left in place longer for safety, commonly referred to as "re-shoring". In this case, king stringers may be used to avoid having to remove and reinstall the shores. As with beams, slight cambering of forms for slabs is sometimes advised to compensate for future sagging under the weight of the concrete and live loads.

Although stress designing is not in the scope of this book, it is worth noting here that slab forms carry a greater live load than other kinds of forms—first the workers and equipment, next the rebar (not yet a dead load), then the wet concrete with accessories, the impact of movement, and sometimes all of these simultaneously. The superintendent cannot be too careful in the construction of slab forms and shoring.

Another method of supporting slab forms uses adjustable horizontal shores that may be rented in various sizes. These shores are placed at such close centers that they serve as joists, and plywood sheathing may be laid directly on them. Their advantages are: (1) speedy erection and removal, (2) economy in lumber, (3) economy in labor, and (4) more open space below for other uses.

Yet another method is to support the slab forms with steel beams, previously erected as a part of the building's basic steel framework. Several different variations are possible. Two of the choices are: (a) using wire beam saddles, and (b) using coil beam saddles. The latter method might be the more expensive of the two, but it has these advantages: (1) it is adjustable for variations in dimensions of wood members, (2) it makes removal of forms easy, and (3) it leaves no projecting metal to rust or to be removed.

Calculating materials quantities for purchase and controlling the cost may proceed as follows, using wood, job-built shoring:

Given: Total area . . . . 10,000 sf
      Total lumber and plywood
      (if all formed at once) . . . . 30,000 bf
      Total number of shores (4" × 4" × 10') . . . . .440 ea.
      Total lumber for bracing, sills, etc. . . . . 2,500 bf
      Reuses of forming materials . . . . .4 ea.
      Reuses of shoring . . . . .5 ea.
      Salvageable material on job site . . . . 1,000 bf

Materials to Purchase:

Lumber & plywood $\dfrac{30,000 \text{ bf}}{4 \text{ uses}}$ = 7,500 bf/use

Add for waste   7,500 bf/use × 4 uses × .10 (waste) = 3,000 bf/use

Shoring $\dfrac{440 \text{ ea.} \times 10 \text{ lf} \times 1.34 \text{ bf/lf.}}{5 \text{ uses}}$ = 1,179 bf/use

Bracing, etc. $\dfrac{2,500 \text{ bf}}{5 \text{ uses}}$ = 500 bf/use

Bracing material waste

     500 bf per use × 5 uses × .10 (waste) = 250 bf/use

Less salvage material   (1,000) bf/use

                                       Total   11,429 bf/use

Cost Control:

Given: Total Budget $35,000
      Time to complete . . . . 31 days
      Average pay scale . . . . $20/Hr.

$$\frac{\$35,000 \text{ budget}}{31 \text{ days} \times 8 \text{ hrs per day} \times \$20 \text{ avg. pay scale}} = 7.06 \text{ workers}$$

In this case, the superintendent plans to complete this work within 31 days with an average of seven workers.

Suspended Stairs

Suspended stairs may be built as soon as possible to provide access to upper floors for workers, materials, and equipment. Reused stairway forming materials lose a lot in waste. 33% might be added for each reuse.

In actual practice, the construction of the different kinds of forms takes place in an overlapping fashion, and thus requires a shorter time span than if each kind had to be completed before another could start. By gathering together into tabular form all the forming examples described in this section, the superintendent can plan the most efficient and economical schedule for formwork, delivery of concrete, and rental of hoisting or pumping equipment. To this end, the superintendent may use a bar chart or CPM network (Chapter 14) and a manning chart (Chapter 15), or an improvised schematic form such as that shown in Figure 28.3.

Summary of Concrete and Formwork

	Uses Each	Purchase Lumber & Plywood bf	Days to Complete Each	Number of Workers Each	Cost Budget $	Concrete Quantity cy
Walls above grade	5	11,000	30	5	24,000	160
Columns	4	1,500	16	3	8,000	19
Beams & girders	2	6,705	21	6	20,400	55
Suspended slabs	4	11,429	31	7	35,000	190
Suspended stairs	2	875	8	2	2,560	7
	—	31,509	106 (gross)	—	89,960	431

Schedule of Formwork and Concrete Placement

Total time = 60 working days

FORMWORK
Walls
Columns
Beams & Girders
Suspended slabs
Stairs
CONCRETE
Placement Cubic yards   20 20 20 24 24 24 35 53 30 30 30 19 19 19 19 21 21 3

Figure 28.3

Finishing of Concrete Surfaces

This process is usually necessary after the forms are stripped. Before constructing the forms, the superintendent will have analyzed the spec requirements and other standards to determine the most suitable forming materials and quality of forming work. The goal is to produce concrete surfaces conforming as nearly as possible to the desired final finish.

The extent of finishing work may range from none at all to a very expensive architectural rubbed and sacked finish. Some of the choices are:

No Finish—is required in the following cases: footing or foundation sides to receive earth backfill; walls or columns to receive veneer or plaster; soffits or beams concealed by suspended ceilings; the inside faces of foundation walls; and the inside faces of parapets.

Form Finish—is acceptable for all surfaces so specified. "Form finish" means to accept whatever surface is found when the forms are removed. Some cases may warrant especially careful formwork and special lining material.

Point and Patch—is not actually a finish, but a touching up of the surface as it is found when the forms are stripped. Filling holes, removing fins, and patching broken corners are all aspects of this process. Pointing and patching is a basic step in all further finishing work.

Rub and Grind—is a required treatment for surfaces to be exposed to public view. This treatment could also be described as a superior pointing and patching job. Filled holes and removed fins are ground flush with the surface planes; an overall uniformity is achieved that is acceptable as it is, painted or unpainted.

Sack—refers to the process resulting in the highest quality "architectural" finish. It is achieved by applying a thin, rubbed-on grout after the grinding work, to conceal stains and cover irregularities in the concrete coloring.

Chapter 29

Precast Concrete Work

Precast concrete is cast in a location other than its final position, then moved, erected, and secured in place. It is used for one or a combination of the following reasons: economy (in cost, time, or working room), special architectural appearance, or when casting in place is impractical or impossible.

Selection of the precast concrete method is most often made by the sponsor's architect/engineer, and the superintendent simply carries it out. However, the precast concrete method is sometimes selected by the superintendent in lieu of cast-in-place (with the sponsor's approval), either as a change order, or when conditions make precasting advisable. Some of the building members that lend themselves to precasting include: walls, columns, beams, girders, spandrel beams, suspended floor and roof slabs and balcony railing. Some site work items may also be precast, such as curbs, wheel bumpers, catchbasins, and manholes.

When performing precast concrete work, the superintendent should give due consideration first to the methods and procedures envisioned by the estimator as the basis for the time and cost budget. Such methods are only guidelines, and the superintendent is encouraged to improve upon them if possible. The following examples are selected to show a variety of conditions that the superintendent might encounter in different precast concrete jobs. It is unlikely that all of them would occur on any single project.

In this section, only field precast concrete tilt slab wall panels will be discussed, but the principles are typical for other kinds of precast concrete members as well. Other than tilt slab construction, most precast members are cast in the controlled conditions of a precasting plant and transported to the site. All types involve formwork, bondbreaker, inserts, concrete, cement finishing, curing, moving, erecting, bracing, and securing. Following is a list of steps approximately in the order they should be taken.

1. Determine the number of similar panels that may be formed and cast in "pancake style" (stacked one on top of another) to save casting slab space.

2. Determine the nearest point to their final locations that panels may be cast, to reduce the time, cost, and risk. The ideal location is that which requires simply tilting up the panels in their final position (no moving).

3. If possible, use the building floor slab as a casting slab. Otherwise, a temporary slab might have to be made, and then demolished later. From the determination of panel numbers and the amount of stacking that can be done, calculate the total area needed for casting slab.

4. Consider which faces are to be finished (faced up and troweled). Locate positions for lifting inserts. These should be well engineered relative to the panel's centers of gravity, so that the panels will experience a minimum of stress and will maintain a stable plane while being moved through the air by crane.

5. Place edge forms, including those needed for window and door openings. For similar or identical panels, the edge forms may be reused a number of times.

6. Spread bondbreaking compound on the casting (or floor) slab to prevent the fresh concrete from adhering to the old, and to ensure an easy, damage-free separation.

7. Lay out and set the necessary miscellaneous inserts, such as bracing inserts, well plates, anchor bolts, reinforcing steel dowels, and window or door frame anchors. These should be carefully rechecked for accuracy, since relocations can be extremely difficult after the concrete is placed.

8. Place all reinforcing steel. This is usually done by a subcontractor.

9. Furnish, place, spread, and vibrate (if required) the concrete. If a large number of panels are cast at the same time, a concrete pump might be preferable to chuting and buggying. Most panels can be screeded off the edge forms, but overwide panels might require intermediate screed chairs.

10. Finish, cure, and protect the panel surfaces. Some specifications might call for special finish operations, such as textured or sandblasted.

11. Before moving wall panels (or any precast member) in place, the superintendent should check to make sure all fastening inserts, or any other means of fastening the precast member, are in place and ready to accept the member.

12. Organize a securing crew to work simultaneously with the crane and riggers. These workers check the panels for plumbness, shimming them at the bottoms as necessary. The crew members also place and secure the braces, and bend and straighten any rebar dowels.

13. Try to make transporting and erecting a smooth, integral operation by casting the panels so near their final position that the lifting crane can carry them, and no other equipment (such as trucks) will be needed. Choose one or more cranes having a total lifting capacity well above the net weight of the heaviest panel. This choice can be assisted through consultation with an equipment rental company. Arrangements may be made for such a company to supply the crane complete with all accessory equipment, operator, oiler, and rigging crew for a given hourly rate.

14. Any required welding is done at this time. Holes remaining at lifting and bracing inserts are patched, the patches blending with the colors and textures of the panels. All other blemishes and imperfections should be removed. Ladders or portable scaffolding may be required.

We will now follow the above outline to demonstrate the construction of a hypothetical precast concrete tilt slab wall paneling job. The superintendent is given the budget in Figure 29.1, along with the project drawings and specifications.

Means Forms
COST ANALYSIS

PROJECT	Example Precast (Tilt-Up) Concrete Panels		
LOCATION		CLASSIFICATION	
TAKE OFF BY PC	QUANTITIES BY	PRICES BY	ARCHITECT

SHEET NO
ESTIMATE NO
DATE 10/31/87
CHECKED BY JM
PRICES BY AT
EXTENSIONS BY AT

Description	Quantity	Unit	Material Unit Cost	Material Total	Installation Unit Cost	Installation Total	Subcontract Unit Cost	Subcontract Total	Total Unit Cost	Total
Casting										
1. Temporary casting slab	5000	S.F.	1.10	5500	0.42	2100				7600
2. Engineering	128	Ea.					13	1664		1664
3. Edge forms	5700	L.F.	.16	912	1.09	6213				7125
4. Bond breaker	16000	S.F.	.10	1600	.029	464				2064
5. Set misc. inserts	1704	Ea.	2.80	4771	1.96	3340				8111
6. Concrete & placing	306	C.Y.	50.90	15575	7.91	2420				17995
7. Finishing & curing	16000	S.F.	.045	720	0.33	5280				6000
Erecting										
8. Crane & riggers	64	Hrs.					172	11008		11008
9. Bottom anchors	256	Ea.	4.00	1024	6.00	1536				2560
10. Brace, level, grout	256	Hrs.	11.70	2995	20.00	5120				8115
Securing										
11. Weld plates	256	Ea.	5.00	1280	8.00	2048				3328
12. Vertical grouting & painting	384	Hrs.	5.00	1920	20.00	7680				9600
				36297		36201		12672		85170

Figure 29.1

1. Determine the number of similar panels that can be cast in stacks, pancake style. The superintendent interprets from the drawings the following quantities:

   Total number of panels . . . . 128 each
   Average size of panels . . . . 125 sf
   Number of panels that can be stacked. . . . . 100 each
   Number of individual panels . . . . 28 each

2. Determine the best location for a casting slab. The superintendent notes that while the estimator allowed for a waste slab of 5,000 square feet (for the uncertainty of the floor slab), the floor slab can, in fact, be used, and most of the budgetted amount can be saved. However, a small part of the budget might have to be spent for floor preparation and protection.

3. Calculate the total area needed for casting. The superintendent finds that 100 panels can be cast in 20 stacks, and the average panel size is 125 square feet. The remaining 28 panels are each unique and must be cast individually. Calculations are as follows:

20 stacks × 125 sf =	2,500 sf	
28 panels × 125 sf =	3,500 sf	
Net area required =	6,000 sf	
Working room =	1,000 sf	
Gross area =	7,000 sf	

   The allowance of 1,000 square feet for working room is for space between the panels and stacks for access by workers. Materials for formwork, bondbreaking, and inserts may now be brought to the job site.

4. Engineer the panels for pickup points. This is usually a free service provided by the manufacturer of the inserts. If needed, experienced engineers are usually available to do this work for a reasonable fee.

   The numbers of pickup points and their exact locations vary with different size panels. These figures are computed, so that lifting inserts may be placed ahead of concrete pours. The inserts receive the pickup cables from the hoisting crane. The lifting shear stresses are distributed for smooth breakaways from the casting slab, and safe plans for swing to the final positions of the inserts.

5. Complete all edge forming, including the surrounds for door, window, vent, and other openings. When there are no projecting rebars, edge forming is simple, as in cast-in-place concrete slabs; when there are projecting bars or keyed edges, the work is more complex and expensive. In our hypothetical example, the forming is greatly simplified because of the re-uses made possible by many similar size panels. Even so, it will be noted that forming labor is the largest single cost item in the budget. The superintendent can control the cost by finding that the budget per panel is:

$$\frac{\$6,123}{128 \text{ panels}} = 47.84, \text{ and } \frac{\$47.84}{\$20/\text{hr.}} = 2.39 \text{ hours.}$$

A crew of two workers must construct forms on an average of one panel in one hour and 12 minutes to keep within the budget. This information is also useful in deciding the number of workers required to complete all the formwork within a given number of scheduled days (see Chapter 15, "Productivity and the Manning Chart").

6. Spread bondbreaking compound on the casting (floor) slab. The gross area to be covered is 16,000 square feet. A good quality of material and a generous coating is good economy in the long run to ensure a clean separation, free from damage to either the panel or floor slab. The control time for application is:

$$\frac{\$464}{128 \text{ panels}} = \$3.625/\text{panel};$$

$$\frac{\$3.625/\text{panel}}{\$20/\text{man-hr}} = .18 \text{ man-hrs per panel}$$

7. Lay out and set all miscellaneous inserts. The estimator has lumped all kinds of inserts together and used average labor and material unit prices. It would be impractical for the superintendent to account for the cost of each one this way. A better method would be to find the average labor cost budget per panel, thus:

$$\frac{\$7,743}{128 \text{ panels}} = \$60.49 \text{ per panel}$$

8. Place all reinforcing steel. The subcontractor should have been informed well ahead of time of the order of panel construction and the time schedule. The superintendent inspects the materials and workmanship of the reinforcing steel subcontractor (see Chapter 16).

9. Furnish, place, spread, and vibrate the concrete. This process is similar to cast-in-place concrete work (see Chapter 27, "Footings, Foundations, and Slabs"). One difference is that the precast work usually involves smaller daily quantities, and somewhat slower placing rates. The budget provides a total of 306 cubic yards of concrete, including a waste allowance. The placing labor averages:

$$\frac{\$1,778}{128 \text{ panels}} = \$13.90 \text{ per panel}.$$

If ten panels are to be poured in one day,

$$\frac{10 \text{ panels} \times \$13.9/\text{panel}}{\$20/\text{per man-hour} \times 8 \text{ hrs}} = .87 \text{ man days} = 1 \text{ man day}$$

This is the maximum number of workers that could be employed to place the concrete assuming a pay scale of $20 per hour. Since you cannot have .87 workers, this calculation shows that you need 1 worker to place the concrete and that he might have between 1 and 2 hours free time that day. The daily pour of concrete would be:

$$\frac{306 \text{ cy}}{128 \text{ panels}} \times 10 \text{ panels/day} = 24 \text{ cy/day}$$

10. Finish, cure, and protect the panel faces. The finishing and curing labor has been budgetted at $4,000. This reduces to:

$$\frac{\$4,000}{128 \text{ panels}} = \$31.25 \text{ per panel},$$

or

$$\frac{\$31.25 \text{ per panel}}{20/\text{hr.}} = 1.56 \text{ man-hours per panel}$$

(assuming $20 per hour pay scale).

If ten panels are to be finished in one day, the number of cement finishers would be:

$$\frac{1.56 \text{ m.h./panel} \times 10 \text{ panels}}{8 \text{ hrs.}} = 1.95 \text{ men} = 2 \text{ men}.$$

This reduces to

$$\frac{125 \text{ sf/panel} \times 10 \text{ panels}}{2 \text{ men}} = 625 \text{ sf per finisher per day.}$$

Common sense and experience tell the superintendent that this production may not be possible, unless superior finishers are available for hire.

11. Choose one or more cranes adequate for the lifting conditions. The superintendent should use the advice of a crane rental company engineer, and may possibly sublet the entire lifting operation. For the crane rental engineer to determine the appropriate size crane, the superintendent will have to provide him the weight of the heaviest object and the longest reach for the crane.

Choose the largest and heaviest precast panel and calculate its weight in tons. For example, if the largest panel is 16 feet wide, ten feet high, and seven inches thick; and the unit weight is 150 pounds per cubic foot, the weight of the panel would be:

$$\frac{16' \times 10' \times 0.58' \times 150 \text{ lbs./c.f.}}{2,000 \text{ lbs./ton}} = 6.96 \text{ tons.}$$

It is possible that the time scheduled for completion might call for the erecting of more than one panel at a time, and therefore more than one crane and rigging crew.

Our sample project allows one half hour per panel for erecting in place. The superintendent reasons that overtime is unavoidable, and will try to achieve better production than this. He will also try to obtain a crane and rigging crew rate of less than the budgetted $172 per hour.

12. Prior to lifting, provide inserts and anchors for bottoms of braces. Inserts are already in the panels for the tops of the braces, but anchorage on the earth, pavement, or floor slab must be provided. The budget allows $1,536 for labor. This reduces to:

$$\frac{\$1,536}{128 \text{ panels}} = \$12 \text{ per panel,}$$

or:

$$\frac{\$12/\text{panel}}{\$20/\text{hr.}} = 0.6 \text{ man-hours per panel}$$

13. Perform the miscellaneous labor associated with the erecting work. It is important that the crane be relieved as soon as possible from holding up the panel, so that it can be moved to more productive work. The crew that levels, braces, and grouts the panel bottoms consists of:

$$\frac{\$5,120}{\$20/\text{hr.}} \text{ divided by 64 hours of erection} = 4 \text{ workers.}$$

14. Secure the panels and accomplish the final finishing and dressing up. At this point, the panels are standing temporarily braced. Columns (if any) are not included in this section, as they are a part of cast-in-place concrete work (see Chapter 28, "Superstructure Concrete Work"). All other connecting is now accomplished as a part of precast work. Welding (if any) might require scaffolding, which is included in the budget under "M/E". The given budget for welding labor is $2,048; assuming $20 per hour for welder pay, the control amount would be:

$$\frac{\$2,048}{\$20/\text{hr.}} = 102.4 \text{ hours,}$$

or

$$\frac{102.4}{128 \text{ panels}} = 0.8 \text{ man-hours per panel.}$$

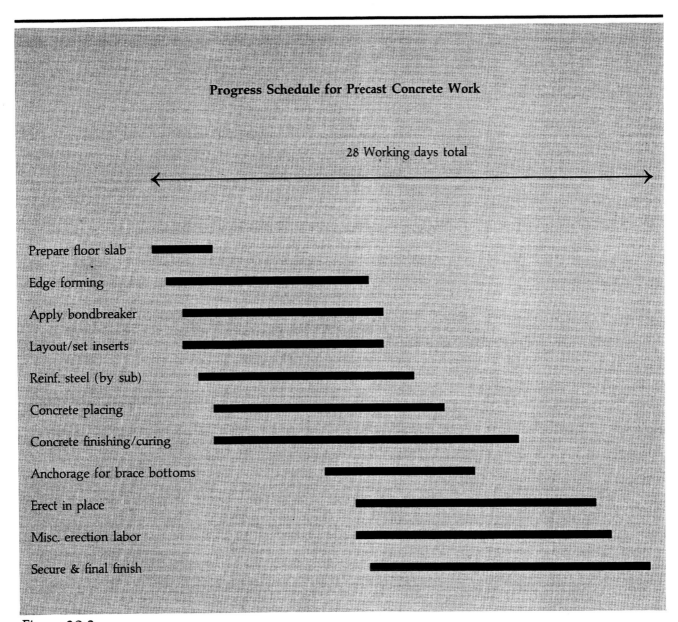

Figure 29.2

Removing of braces, vertical grouting, and pointing and patching of insert holes and other blemishes, are budgetted for labor at $7,680, which reduces to the control amount of:

$$\frac{\$7,680}{128 \text{ panels}} = \$60 \text{ per panel},$$

or

$$\frac{60/\text{panel}}{\$20/\text{hr.}} = 3 \text{ man-hours per panel.}$$

This completes our hypothetical precast concrete project, except for the "postmortem", which cannot be thoroughly conducted until all of the associated labor, material, equipment, and subcontract cost items have been documented.

Finally, the superintendent must determine whether or not the precast concrete work can be accomplished within the time span in the project progress schedule. A bar graph can, as shown in Figure 29.2, can be a useful tool for answering this question.

Let us assume that the progress schedule limits the precast work to 30 days. Our bar graph, showing the overlapping of operations, indicates that an average production of ten panels per day could complete the work in 28 days—assuming no bad weather or mishaps.

Chapter 30

Rough Carpentry

Today's medium size contracting company is most likely to be involved in fireproof or semi-fireproof construction such as industrial, commercial, and military. This type of construction uses minimal lumber for permanent members. Most of the carpentry work is temporary and incidental, such as formwork for concrete. Still, a well-rounded superintendent should have proficiency in carpentry work for such items as wood partitions, floors, and ceilings. This section will present the kind of overview that a superintendent needs—the experience required to order materials, schedule the work, organize the crews, and oversee the work.

The trade classification, "rough carpentry", applies to wood material, predominantly structural in purpose, and eventually concealed from view (as opposed to "finish carpentry", which must meet the critical eye). The essential rough hardware (nails, bolts, joist hangers, etc.) is an integral part of the material and work that falls under this heading.

The superintendent may receive both labor budgets and material lists from the company's estimating/bidding department. These budgets and lists are not intended as rigid standards. The superindent is expected to check them carefully for accuracy and to improve upon them if possible.

An example of rough carpentry work is used in this section. The superintendent has received from the company estimator the following budget (Figure 30.1), lumber and plywood list (Figure 30.2), and rough hardware list (Figure 30.3). The supplier has been previously selected and awaits only the formal order to deliver.

When a lumber and hardware list is made by an experienced quantity surveyor (estimator), it is possible that the materials as listed can be ordered without further adjustment. Even so, the superintendent should take the time to check the most important or critical items, such as premium lengths of rafters, joists, beams, and other heavy timbers; and the classifications of members specified as "select structural", "kiln dried", "pressure treated", or "rough sawn".

Understandably, the estimator is forced to make a number of assumptions when it comes to calculating and ordering materials. Nevertheless, he tries to be on the "safe side" by collecting as much relevant information as possible. By checking the material list, the superintendent might be able to make substantial savings. Ambiguities can be resolved by conferences with the sponsor's representative.

Example of Rough Carpentry Price-out

Description		Quantity	l	m	L	M	T
All lumber per list & quote		60,502 bf	—	ls	—	29,948	29,948
All plywood per list & quote		24,960 sf	—	ls	—	10,982	10,982
All hardware per list & quote		ls	—	ls	—	2,688	2,688
Labor-roof facia	3x10	1,000 bf	.50	—	500	—	500
Labor-roof ledgers	3x10	800 bf	.615	—	492	—	492
Labor-roof beams	6x12 4x10	800 bf	.444	—	355	—	355
Labor-roof joists	2x10	10,020 bf	.372	—	3,727	—	3,727
Labor-roof blocking	2x10	1,000 bf	.543	—	543	—	543
Labor roof X bridging	2x3	750 bf	1.127	—	845	—	845
Labor-floor ledgers	3x14	1,400 bf	.529	—	741	—	741
Labor-floor beams	6x14 8x8	1,072 bf	.362	—	388	—	388
Labor-floor joists	2x14	14,000 bf	.329	—	4,606	—	4,606
Labor-floor blocking	2x14	1,400 bf	.529	—	741	—	741
Labor floor X bridging	2x3	750 bf	1.073	—	805	—	805
Labor-ceiling ledgers	2x6	800 bf	.50	—	400	—	400
Labor-ceiling joists	2x6	7,000 bf	.394	—	2,758	—	2,758
Labor-ceiling blocking	2x6	1,600 bf	.50	—	800	—	800
Labor-wall sills, bolted	2x4	1,333 bf	.50	—	667	—	667
Labor-wall studding	2x4	10,666 bf	.358	—	3,818	—	3,818
Labor-wall blocking	2x4	1,333 bf	.50	—	667	—	667
Labor-wall top plates	2x4	2,666 bf	.429	—	1,144	—	1,144
Labor-wall bracing	1x4	112 bf	1.073	—	120	—	120
Labor-misc. stripping	1x3	400 bf	1.073	—	429	—	429
Labor-misc. grounds	1x2	200 bf	1.662	—	332	—	332
Labor-plywood roof sheathing	5/8"	10,240 sf	.312	—	3,195	—	3,195
Labor-plywood wall sheathing	1/2"	2,560 sf	.332	—	850	—	850
Labor-plywood floor sheathing	5/8"	10,240 sf	.312	—	3,195	—	3,195
Labor-plywood shear panels	1/2"	1,920 sf	.351	—	674	—	674
Equipment hoisting (48 hrs. @ 70/hr)		70,000	—	.048	—	3,360	3,360
Total					32,792	46,978	79,770

Figure 30.1

Description		Size	No. Pcs.	Length	Bd. Ft.
Example of Lumber List					
Roof fascia	const grd. SA	3x10	Ran	400	1,000
Roof ledgers	const "	3x10	Ran	320	800
Roof beams	struc "	6x12	4	20	480
Roof beams	struc "	4x10	6	16	320
Roof joists	const "	2x10	100	20	3,340
Roof joists	const "	2x10	250	16	6,680
Roof blocking	std "	2x10	Ran	600	1,000
Roof X bridging	std "	2x3	Ran	1,500	750
Floor ledgers	const "	3x14	Ran	400	1,400
Floor beams	struc "	6x14	4	20	560
Floor beams	struc "	8x8	6	16	512
Floor joists	const "	2x14	100	20	4,667
Floor joists	const "	2x14	250	16	9,333
Floor blocking	std "	2x14	Ran	600	1,400
Floor X bridging	std "	2x3	Ran	1,500	750
Ceiling ledgers	const "	2x6	Ran	800	800
Ceiling joists	const "	2x6	350	20	7,000
Ceiling blocking	std "	2x6	Ran	1,600	1,600
Wall sills	const "PT	2x4	Ran	2,000	1,333
Wall studding	const "	2x4	2000	8	10,666
Wall blocking	std "	2x4	Ran	2,000	1,333
Wall top plates	const "	2x4	Ran	4,000	2,666
Wall bracing	const "	1x4	24	14	112
Miscellaneous stripping std "		1x3	Ran	1,600	400
Miscellaneous grounds std "		1x2	Ran	1,800	200
Miscellaneous waste std "		2x4	Ran	2,100	1,400
Total					60,502
Plywood					
roof sheathing 5/8"		4'x8'		320	10,240
wall sheathing 1/2"		4'x8'		80	2,560
shear panels 1/2"		4'x8'		60	1,920
floor sheathing 5/8"		4'x8'		320	10,240
Total					24,960

Notes: All lumber is Douglas Fir s4s.
All plywood is Douglas Fir CDX Struc. I. PT = pressure treated.
SA = select for appearance. RAN = random length.

Figure 30.2

Absolutely accurate quantities of rough hardware stock items are not crucial since shortages can quickly be resolved with a purchase, and excess can be either returned to the supplier or kept in storage for use in a future project. Unusual (custom design) rough hardware, shortages of which could cause delays in job progress, or excesses of which are not apt to be used in future projects, should be carefully checked for both sizes and quantities.

Lumber and plywood should not be ordered for delivery too early, as storage room on the job site is limited, and the risk of damage and theft is great.

Example of Rough Hardware List		
Description	**Size**	**Quantity**
Joist hangers	2 x 10	350 ea
Joist hangers	2 x 14	350 ea
Tie straps	1/4 x 2 x 30	28 ea
Post caps	4 x 4	18 ea
Post bases	4 x 4	18 ea
Anchor bolts	5/8 x 12	120 ea
Anchor bolts	1/2 x 10	80 ea
Machine bolts	1/2 x 8	60 ea
Machine bolts	5/8 x 10	100 ea
Lag bolts	3/8 x 6	75 ea
Square washers	1/2	180 ea
Square washers	5/8	220 ea
Nails, box	8d	600 lbs
Nails, box	16d	300 lbs

Figure 30.3

Upon delivery, the superintendent should provide as much protection for the material as possible, and see that it is divided into identifiable stacks, separating roof, floor, ceiling, wall, and other categories. Inferior pieces should be culled out for replacement. Quantities, dates of delivery, and conditions should be noted in the job log book (see Chapter 10, "Daily Reports, Diaries, Logs, and Recordkeeping"). At this point, appropriate security measures should be provided for the lumber (see Chapter 7, "Project Security").

Rough hardware should be stored and locked in a weatherproof shed (see Chapter 6 for further discussion on planning the working area.)

The project progress schedule allots a certain time span for the starting and finishing of rough carpentry work. It can be expected that productivity will occur in peaks and valleys, with variable sizes of worker crews.

A bar graph, like the one used for the precast concrete work (Chapter 29, Figure 29.2), can be drawn up to determine the number of workers required to meet the time schedule. If the superintendent does not feel the need to go into that much detail, a simple alternative is the formula used often throughout this book, namely: cost budget divided by the average worker pay scale, divided again by the number of allotted working days.

In our example, the labor budget is $32,792. We will use $20 per hour as the average base pay, and if the scheduled time to complete is 90 days, we have:

$$\frac{\$32,792}{\$20/hr.} \div 90 \text{ days} = 18.22$$

If the superintendent can accomplish all rough carpentry in 90 days with an average of 18 workers, the cost will not exceed the budget.

The superintendent gives only secondary interest to material cost control. Once quantities and prices are determined, the material cost remains more or less stable. By contrast, labor (and equipment), being dependent on many production variables, rightfully claims the majority of the superintendent's attention.

Chapter 31

Managing Subcontract Trades

Previous chapters in this section cover the scope of work in which the superintendent is most involved. These are trades that require the superintendent to check the time and production of both men and machines. This chapter reviews the scope of work that is generally awarded to subcontractors for a negotiated lump sum. For these subcontracted trades the burden of checking time and production for men and machinery becomes the responsibility of the subcontractors. An exception to this rule is work done on a time and material basis. In such cases, the superintendent must monitor the work items.

The bid package (see Chapter 9, "Introduction to the Project") received by the superintendent at the start of the project identifies the work areas to be accomplished by the superintendent, those that are the joint responsibility of the superintendent and the subcontractors, and those that are to be handled by the subcontractors alone. This chapter concerns the superintendent's relationship to some of the most common subtrades. As it is impossible to list every trade that a superintendent might encounter, coverage is confined to the list below.

Reinforcing Steel

Reinforcing steel and mesh work is almost always sublet; but certain associated duties remain the superintendent's personal responsibility; proper earth subgrades, basic layouts, and templates are in this category. Scaffolding, hoisting and welding are other items which may not always be part of a rebar subcontractor's contract (N.I.C.). Such questions of responsibility are usually determined in the earlier bidding phase and are called out in the subcontract, thus preventing arguments in the field. The estimator should provide a cost budget for these N.I.C. items.

Sometimes the rebar subcontractor, by agreement during or after the bidding phase, may supply certain materials cut, bent, or prefabricated for the superintendent to install using his own company's crews. In some cases, a project may require such minor or diverse items that a subcontractor would either not be interested or charge excessively. The superintendent might then calculate and purchase the materials—either "loose-and-long" or cut and bent—and supervise their installation by company crews. A knowledge of conventional materials, accessories, rules of overlapping, and techniques of securing is necessary. In the case of a major reinforced concrete project, the superintendent will need to provide the subcontractor with ample storing and fabricating space (see Chapter 6, "Planning the Working Area").

Since rebar is often needed very early in a project, the superintendent may have to urge the subcontractor to submit shop drawings, and the sponsor's engineer to speed the approval. As further motivation for speed in this matter, an allotted time span is shown on the project progress schedule (see Chapter 14, "Time Controls").

Masonry

Masonry, like rebar, is almost always sublet. The superintendent receives a subcontract agreement to initiate the work and supervise it through to completion. Some associated items may be N.I.C. and remain the responsibility of the superintendent. Examples are the shoring of wall openings, hoisting, and the furnishing of scaffolding or reinforcing steel. A number of mutual agreements and decisions may have to be made in the field between superintendent and subcontractor. Some examples are listed below.

- The starting point of the work
- Purposeful delays between courses
- Heights of lifts
- Quantity of units placed per day, or week
- Rates and methods of grouting
- Locations of dowels, bolts, and other inserts
- Responsibility for bond beams, reveals, and caulking
- Temporary shoring for wind
- Curing, cleaning, and patching
- Use of scaffolding

As with reinforcing steel and other major subtrades, the superintendent may have to provide storage space for the masonry subcontractor.

In cases where it is not feasible to sublet masonry work, the superintendent might calculate quantities of materials and purchase them, rent mixing machines, hire masons, and supervise the work. Since masonry work is closely related to the routine trades typically performed by the general contractor, the superintendent should have little difficulty carrying out simple, standard masonry items when expedient.

Miscellaneous Metal

Miscellaneous metal may be subdivided into (1) structural steel, (2) miscellaneous metal, and (3) manufactured stock size items. Ideally, a superintendent has one subcontract that includes all three divisions. Unfortunately, it is usually not that simple. Structural steel work is almost always sublet. Within this division, the superintendent's concerns are shop drawings, inspection of weldings, safety, scheduling, and working room. As a rule, he must anticipate and provide for anchorages for such items as columns to footings and beams to walls, by accurate placement of bolts and inserts in the wet concrete or masonry. Even though the subcontractor is usually responsible for correct layout measurements and elevations, the superintendent is wise to carefully check them all before allowing other contingent trades to proceed (see Chapter 19—"Safety Precautions and Quality Controls, Inspections and Tests").

Miscellaneous metal is generally nonstructural and is processed or fabricated off the job site to custom design specifications. Examples are tailings, ladders and catchbasin covers. Subcontractors supply miscellaneous materials based on a previous agreement and schedule, installing certain items and delivering others for the superintendent to install at his own convenience (within a cost budget provided by the company estimator). Other items in this category are manhole covers, channel door frames, sleeves for railings, and stair nosings.

The third division, stock metal items, may also be furnished and possibly installed by the subcontractor. The super can also purchase such material directly from steel supply houses, hire the appropriate trade workers, and supervise the installation. Stock items include door and window opening lintels, truss members, and angle door thresholds.

Millwork

Millwork is usually furnished F.O.B. jobsite by subcontract or purchase order. When there is a lot of custom work, such as cabinets, and only a small percentage of stock items, the subcontract might be all-inclusive, in which case the superintendent has no extra items to purchase. He does have the following concerns:

- Scheduling the order of material deliveries to the site (certain items will arrive ahead of others).
- Receiving deliveries, checking them for correctness, distributing them to rooms or temporary storage, and protecting them from damage.
- Acquiring all hardware, rough and finish, required for the installation.
- Supervising the installation crews.
- Coordinating associated trades, such as plumbing, electrical, installation of laminated plastic tops, and painting.

In some projects, the millwork consists almost entirely of manufacturer's stock shapes. In such cases, the superintendent is required to make an accurate takeoff (or use that previously made by the estimator), obtain prices, make purchases, and schedule delivery dates.

The millwork supplier's contract may sometimes include the installation of at least a portion of the cabinets and materials. Those items left to be installed by the superintendent have cost budget limitations. The budget covers all "finish carpentry" labor as well as millwork. All wooden material is furnished by the millwork subcontractor and installed by workers under the supervision of foremen and the superintendent. Metal doors and frames and finish hardware are supplied under other purchase orders, but installed under this heading along with the millwork.

It is important that the superintendent keep a record of the actual cost of each item for purposes of future cost estimating (see Chapter 13, "Cost Controls").

Sheet Metal

Sheet metal work, if extensive and complex, is usually sublet. Shop drawings may be required of the subcontractor for items that are custom (shop) made, and catalog data may be needed for stock manufacture items, such as wall louvers. The superintendent's main concern is preparation for the installation. Some items, such as roof counter-flashing, require reglets embedded into concrete or masonry parapets. The superintendent must make certain that these and other inserts are placed, and placed accurately.

Windows and Metal Door Frames

Windows and metal door frames concern the superintendent early in the project when sponsor's approval is needed. Since a certain amount of time is required for the manufacturing, the superintendent cannot permit any waste of time in the processing. In some cases, construction conditions require that door or window frames be installed before walls are completed, or inserts may have to be set in place before concrete or masonry walls can be finished. It is very important that exact rough opening dimensions are available for window and door frames that will be installed after walls are built.

Usually window subcontracts include installation by the subcontractor, with or without glazing. Subcontracts should outline the responsibility for caulking and sealing. If glazing is included by the subcontractor, responsibility for initial cleaning, and for replacement (in cases of breakage) should also be previously determined.

Finish Hardware

As a rule, finish hardware is not sublet. Competitive bids are received, and a purchase order is written in order to avoid itemizing. Ideally, the supplier is required to furnish all finish hardware. In this way, the superintendent need not be burdened with shortages and extra costs.

Early in the construction process, the supplier submits a hardware list. At that time, the superintendent, the sponsor and others have the opportunity to make sure it is complete and correct. The superintendent may then request delivery on a certain specified time schedule. The hardware should be stored in a weatherproof, locked and secured room or shed to which only the supervising foremen have access. Keys should be tagged for identification, and retained safely for future surrender to the sponsor.

Painting

Painting is almost always sublet. Except for obtaining samples and color selections from the sponsor, and scheduling the various items to be painted, the superintendent is not greatly involved. Quality control is a major concern, since painting is both protective and cosmetic. The superintendent needs to maintain a systematic inspection to assure that the specified materials are used, proper surface preparations are made, and the correct number of coats are applied.

In some projects such as engineering and underground utility jobs that involve only a small amount of painting, the superintendent might carry out all the painting work without subletting. In such cases, the specifications are not difficult to meet.

Drywall, Metal Studding, Furring, Lathing and Plastering

Drywall, metal studding, furring, lathing, and plastering must be protected from wet weather. Layouts for partitions, doors, and so forth cannot be left entirely to the subcontractor; the superintendent should check all layouts personally. Before drywall or plaster is applied, all in-wall piping and electrical materials must be in place. The superintendent must obtain affirmation from the plumbing, electrical, and other subcontractors that their work is completed in the areas to be covered. If such an affirmation cannot be obtained, the super may notify the sub in writing that certain designated walls will be covered on a specific date, unless the sub requests postponement. The responsibility for cutting and patching is then placed on the subcontractor.

Still another item must be checked before wall covering is applied. The superintendent must make sure that all necessary blocking is in place in the studding for attachment of such items as cabinets and plumbing fixtures. In some projects, blocking is part of a big layout job, and is not an insignificant cost.

Resilient Flooring

Resilient flooring concerns the superintendent in the following ways:

- Proper preparation of subfloor surfaces is a common point of contention, even when a subcontract makes the subcontractor totally responsible. Preparation should be a cooperative matter. The superintendent should make sure that subfloor surfaces are smooth and level and protected against damage from construction crews and equipment. Rough patching and preparation should be the superintendent's responsibility.
- Dryness of the subfloor is often left to the judgment of the subcontractor. The superintendent should be aware of the state of dryness and other conditions of subfloor preparation in order to restrain an overly eager subcontractor, or to prompt a malingerer (see Chapter 2, "Responsibilities").
- Quality of workmanship is important, as resilient flooring has high visibility.
- Cleaning, waxing and protecting of finish flooring, usually part of the subcontract, involves the superintendent in problems of traffic, barricades, and coverings.

Suspended Ceilings

Suspended ceilings, whether for drywall, plaster, or acoustic tile, require preparatory layout for hangers and perimeter angle supports. Even if this work is the subcontractor's responsibility, the superintendent must make sure it is done, as well as insuring that ceiling inserts are properly spaced, and ceiling clear-heights are correct.

Before finish material is placed on ceiling framework or furring, the superintendent must be sure that all other work, (including ducts, electrical, and plumbing) is completed and that access panels have been provided to certain areas for future mechanical and electrical repairs.

Building Specialties

Building specialties are usually arranged (through purchase orders) for delivery to the job site upon request from the superintendent. Many specialties must be attached to the structure, requiring the superintendent to anticipate exact locations and to provide supports, inserts, etc. In some cases, information must first be obtained from the supplier regarding sizes and spacings. Examples of building specialties and their requirements are listed below:

- Flagpoles (need footings and sleeves)
- Building name letters (concrete inserts)
- Aluminum handrails (sleeves)
- Toilet partitions (embedded bolts)
- Bathroom accessories (inserts)
- Venetian blinds (blocking)

Mechanical Work

Mechanical work is a general term that covers several closely related trades. Included in this category are underground water, sewer, air and steam lines, interior plumbing, heating, ventilating and air conditioning. Mechanical work can be one of the major trades in a project, and is almost always sublet. A superintendent for the general contractor need not be an expert in the fine details of the mechanical trades; his expertise lies in the coordination of the mechanical work along with the other trades.

Mechanical work is often one of the first to begin in a project, and one of the last to finish. Some of the superintendent's typical duties regarding mechanical work are listed below:

- Expedite the approvals of materials, samples, catalog data, etc.
- Provide basic surveys and layouts for pipelines
- Check subcontractor's shop drawings and floor plan layouts
- Plan and schedule the sub's trenching and shoring work for safety, and for minimum interference with other operations.
- Provide the subcontractor with yard space for material and equipment storage, and for fabricating the work. As a major subcontractor, the mechanical contractor may also need room for a complete, separate job office and storage shed.
- Carefully study the various floor plans (first, second, third . . . etc.) for vertical alignment of all pipes and fixtures. Faulty design drawings should be corrected at this time, either by the sponsor or by as-built methods, to prevent costly construction changes later.
- Provide access to under-floor areas and wall lines so that the subcontractor can rough-in plumbing and ductwork before the placement of concrete slabs. Expedite all underfloor work and obtain inspection approvals to minimize delay in follow-up structural work.
- Coordinate all mechanical work as it proceeds above ground with the work of all other structural and architectural trades.
- Supervise all mechanical testing, and collect the test results for delivery to the sponsor at the project's completion.
- See that all exposed parts of the mechanical work are properly painted and aesthetically acceptable.

Electrical Work

Electrical work is a major trade that is almost always sublet. The superintendent's duties and responsibilities regarding electrical subcontract work parallel those for mechanical work, namely:

- Expedite material approvals.
- Furnish basic layouts.
- Check shop drawings.
- Plan and schedule the order of procedure.
- Provide yard space.
- Provide access for rough-ins.
- Coordinate with other trades.
- Supervise operational testing.

Repair and Remodeling

If a project involves this kind of work, the superintendent should begin by determining its size and complexity. If the work is extensive, the superintendent might hire an experienced alterations foreman to carry the burden of detailing, supervising and scheduling. In this way, the superintendent remains free to oversee, inspect, and coordinate the other subcontracted work.

Some alteration work items can be done concurrently, while others are contingent on a previous or following activity. The various sucontractors rely on the superintendent for directions regarding what to do, and when (see Chapter 2, "Responsibilities").

While most repair and remodeling items are independent of each other and, theoretically, could be accomplished at the same time, there may be other factors that require a time separation. Some of these factors might be: (1) prefabrication and delivery of materials, (2) economy in numbers of crews, and optimum employment of workers, (3) consideration for subcontractor's performance economics, (4) strategic use of space, (5) structural and safety considerations, and (6) security and weather considerations.

Daily person-to-person contact is necessary between the superintendent, the alterations foreman, and the subcontractors. Each contributor should show the specifics necessary for planning his materials and crews at least one or two days in advance of his performance date.

Chapter 32

Summary and Conclusions

This book covers various aspects of the project superintendent's work in a general contracting company. This final chapter is a summary of the most prominent concerns of this job. The superintendent takes full charge of an individual construction project, beginning with the conditions set during the earlier bidding procedures. The superintendent accepts the economies in the bid package and attempts to accomplish the entire project within the given time and cost limitations, and with the maximum quality of material and workmanship.

A unique combination of personality and experience is necessary in a competent superintendent. Important qualities include construction know-how, motivation, leadership, decision-making ability, public relations talent, as well as visual and organizational ability.

The superintendent is generally given a great deal of autonomy in managing a project. However, he is also expected by his company to be single-minded concerning the project and to bring in a "winner".

The superintendent does not have to be an expert in all trades. He does need to be proficient in the supervision and coordination of a large team of experts. Like a police officer directing traffic, a superintendent must skillfully control and direct the complex "flow of traffic" that interacts in a construction project. Part of this responsibility involves scheduling various trade groups, setting standards of speed and safety, keeping records, and regulating the pace according to weather conditions. He is also responsible for maintaining security, protecting the environment, witnessing, and sometimes adjudicating in clashes and disputes.

The superintendent's job begins when he receives the bid package and determines the scope of responsibilities of all suppliers and subcontractors. Part of this process is eliminating overlaps, and bridging any gaps. The superintendent plans and lays out the working area and the project site for the most effective use and traffic movement. He arranges the erection of barricades and fences as required or advisable, and is also responsible for erecting an office structure and furnishing it with equipment and staff. The superintendent follows the guidelines for all general conditions expenditures, including such items as office, storage sheds, toilets, telephones, water and electric service, fences, signs, small tools, barricades, clerical salaries, and utility charges.

The superintendent arranges an appropriate progress schedule to control the performances of subcontractors and workers, ensuring that the project will be completed within the given time span. Time records are maintained for the completion of individual work items. These figures can then be compared to the estimated time and costs. This process provides a second basis for future scheduling and estimating. It also tends to control costs during the project and keep them within the budget limitations.

The superintendent attempts to save on the cost of materials, but places more emphasis on the saving of labor (saving means spending less than the estimated budget amounts). A good way to save on labor is to determine how many man-hours are allowed by the cost budget, and then try to avoid exceeding that number. Other ways to control and save on labor are:

- Extensive planning, detailing and laying out work.
- Arranging for optimum size crews.
- Planning crews for full days or half days of work.
- Having small jobs available to assign to surplus workers, thereby avoiding "standing-around-time".
- Using tools and equipment to reduce hand labor.
- Offering bonuses and incentives to workers.

The superintendent receives an estimated labor cost budget. Drawing up a manning chart is a good way to make sure that the budget is not exceeded. While a budget might prove—in the long run—to have been estimated too low, a manning chart can help to avoid waste in labor costs.

The superintendent is torn between two opposing demands: to minimize time and cost, and to maximize quality. Specifications provide minimum standards of materials and workmanship acceptable to the sponsor. Both cost and quality are finally a matter of some compromise.

The superintendent is committed by labor relations laws to certain pay scales, fringe benefits, minority hiring practices, and other hiring/firing regulations. Construction work proceeds in such a way as to create a flow of "transient" workers. When a superintendent discovers superior individuals and crews, he attempts to retain them as semi-permanent company employees.

Almost without exception, changes are made to the original project drawings and specifications. The superintendent is responsible for analyzing the extent of these changes, initiating the cost estimating procedure, and participating in the final negotiations with the sponsor.

A fundamental aim of the superintendent is to improve the productivity of both company workers and subcontractors. The first can save money, and both can save time. Up to a point, efforts to save time and cost can be fruitful; beyond that point, diminishing returns make the effort less worthwhile.

Relations with sponsors are a practical matter, and depend to a great extent, on the sponsor's attitude. The superintendent's best approach is to follow guidelines set by the company manager. In general, project cost estimates reflect the company's past experiences with particular sponsors. Previous conflicts of interest between contractors and sponsors led to the establishment of quality standards and tolerances, generally enforced by the courts, should litigation occur. The superintendent should be familiar with these standards. Monthly progress payment requests are another responsibility of the superintendent. These requests are based on the various material quantities installed, or percentages of work items completed. It is important to the company that maximum payments are received. On the other hand, excessive requests can cause delays in payments—to the inconvenience or hardship of some subcontractors. It may be necessary for the super to spend a lot of time researching, computing, and negotiating with subcontractors and sponsor's agents to accomplish a mutually agreeable payment request.

The superintendent may at times find it economical and convenient to sublet certain items to specialists, rather than have the work done by company employees. Some common examples involve equipment work (e.g., trenching, hoisting, surveying, dewatering, concrete finishing, and caulking). Such work might be arranged using a work order or a simplified form of contract.

One of the superintendent's regular duties is enforcement of the terms of subcontracts. There are many ways in which subcontractors can fall short, and foresight is the best way to avoid these shortcomings. However, when a serious problem arises, the superintendent should notify the subcontractor in writing of the possible consequences. The next step is to hand the problem to the company manager for more drastic measures.

One common method to expedite disputed work is to have it done by company employees and backcharge the unwilling subcontractor by deducting a portion of his next payment. This is a drastic step that should be used only after other approaches have failed (see Chapter 18, "Process Payment Requests and Backcharging").

During construction, some details inevitably vary from the original plans (e.g., routing of pipes and electrical conduits). Many of these variations do not affect cost, and are simply approved by the sponsor's field representative. The superintendent keeps a record of such changes on "as-built" drawings. The sponsor's representative should initial these changes to avoid the possibility of future disputes. Items not detailed on the contract drawings might require on-site shop drawings or simple sketches.

The superintendent is responsible for all the basic layout work; but even in such cases where subcontractors do the final layout, the superintendent should still check the dimensions and elevations carefully. Having to make corrections later on can affect the overall progress and quality of the project.

Those trades that have been sublet (usually a majority of the project) require a different approach from the superintendent than those trades that are accomplished by company employees. Subcontractors bear the full responsiblity for their own materials and workmanship and therefore deserve a high degree of autonomy. The superintendent's concerns regarding subcontractors can be covered in three divisions (1) prepare the way for each subcontractor to start on his scheduled date, (2) keep the subcontractors moving at their scheduled paces, and (3) obtain an acceptable quality of materials and workmanship at the completion of each trade.

The traditional superintendent supervised the work of his own company's employees and monitored the performance of the subcontractors who accomplished the remaining one third of the work. Today's superintendent has become a field project manager, directing and coordinating a much larger percentage of subcontractors and overseeing the administration of contracts, change orders, and purchase orders. The superintendent's role is to manage the project from start to finish—to successfully fulfill and complete the contract within the requirements of cost, time, and quality.

Part IV

Appendix

Layout Techniques

Plane Geometry (dealing with two dimensional shapes) has a wide application throughout the construction industry. It is used for almost all layout work.

The most commonly used geometric shape is the 3-4-5 right triangle, shown in the figure below. The side opposite the 90 degree angle is always the longest, and the lengths of the sides always remain in the same proportion.

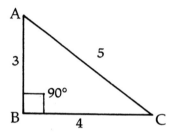

The triangle does not have to be just 3′ by 4′ by 5′. It can be adjusted to a smaller or larger triangle by multiplying or dividing all three sides by the same number. For example, all of the triangles in the figure below are 3-4-5 right triangles.

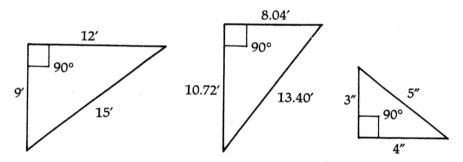

Another common geometric principle often used when one inclusive angle is 90 degrees and the diagonals are equal in a four-sided shape is that all four angles in that shape are 90 degrees. With these two principles it is possible to accomplish most construction layouts.

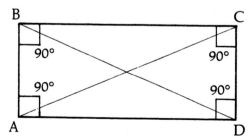
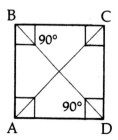

As an example, a rectangular slab must be laid out parallel to and 5' distant from a property line and checked for square. The slab measures 12' × 20'.

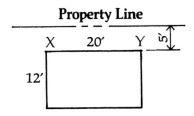

Step 1: Points P and L are marked out 20'–0" apart along the property line.

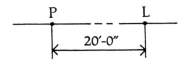

Step 2: Using the 3-4-5 right triangle; points X and Y are staked. First, point X is determined by holding one tape measure at point P and pulling the tape measure out until 15' shows. Another tape measure is held at point Z and pulled out until 25' shows. The zero ends of the two tapes are secured at points P and Z, respectively, and are crossed until 20' and 25' overlap. This establishes a line PX perpendicular (at right angles or 90 degrees) to the property line with a length of 20'.

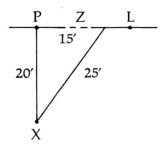

Point Y is established using the same method.

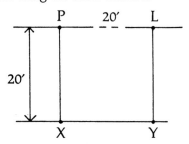

A 20' length was chosen because once the lines PX and LY are established, other lengths can be measured along each line. The closer side of the slab can be established by measuring 5' off the property line along both lines PX and LY. Now points A and B can be marked, establishing line AB (the edge of the slab).

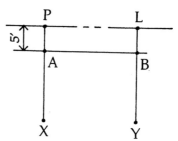

Two additional measurements can be made along PX and LY, to mark the other side of the slab, points C and D. This time, 17' from the property line or 12' from points A and B.

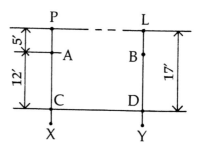

Points ABCD have marked the proposed 12' × 20' slab 5' from the properly line. The slab is square if both diagonals, when measured, are exactly the same length.

Although the 3-4-5 triangle is a very easy tool to use, it is not the only one available for layout. Other layouts may require the use of another geometric principle, the Pythagorean theorem. Simply stated, it says that in a right triangle, the square of the length of the side opposite the 90 degree angle equals the sum of the squares of the other two sides.

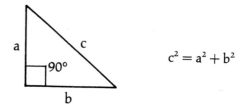

$$c^2 = a^2 + b^2$$

Using the same example, points ABCD could have been established using this method.

Step 1: A 20'–0" is marked out along the existing property line.

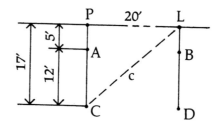

Step 2: Using the Pythagorean theorem:

$$c^2 = 20^2 + 17^2$$
$$c^2 = 689$$
$$c = \sqrt{689}$$
$$c = 26.25'$$

Now two tape measures can be used to establish point C. The first tape is secured at point P and pulled out to 17'. A second tapemeasure is secured at point L and then pulled out to 26.25'. The intersection of the two tapes designates point C. Point D can be located using the same method.

Points A and B can then be established by measuring 5' off the property line along lines PC and LD. Checking for square is done in the same way as in the previous example, by measuring both diagonals to make sure they are the same length.

Additional geometric shapes and relative formulas for areas and volumes appear on the following pages. This information may be useful to the superintendent for layout work.

Area Calculations

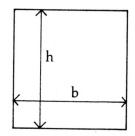

Square   Area $= h \times b$

Rectangle   Area $= h \times b$

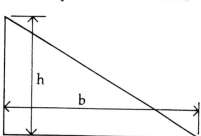

Triangle   Area $= \dfrac{h \times b}{2}$

Parallelogram   Area $= h \times b$

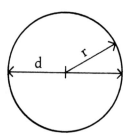

Circle   Area $= \dfrac{\pi d^2}{4} = \pi r^2$

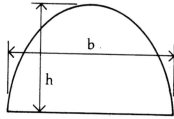

Parabola   Area $= \dfrac{2hb}{3}$

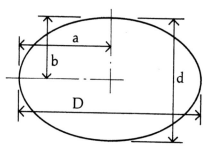

Elipse   Area $= 0.7854Dd$

200

Volume Calculations

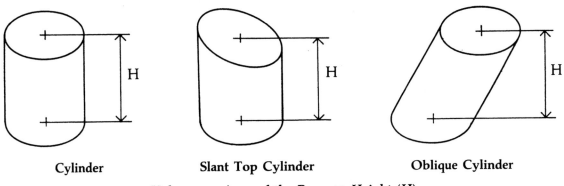

Cylinder   Slant Top Cylinder   Oblique Cylinder

Volume = Area of the Base × Height (H)

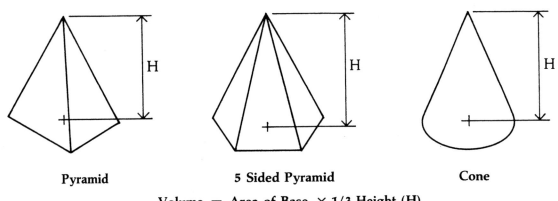

Pyramid   5 Sided Pyramid   Cone

Volume = Area of Base × 1/3 Height (H)

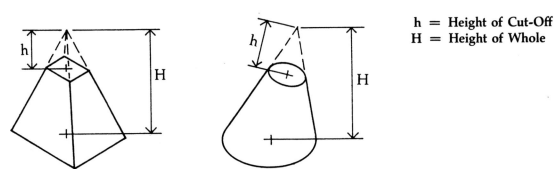

h = Height of Cut-Off
H = Height of Whole

Truncated Pyramid   Truncated Oblique Cone

Volume = Volume of the Whole Solid Less Volume of the Portion Cut Off

Geometric Properties

1. Divide a line into equal parts. Example: divide AB into 9 parts.

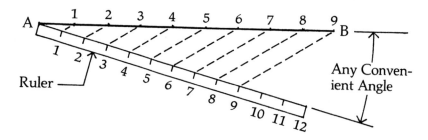

Draw parallel lines 9-9, 8-8, . . . , 1-1

2. Drop a perpendicular to a line AB. From a center at any point E, swing an arc that cuts line AB at C and D. From centers at C and D, swing short arcs intersecting at F. Draw line EF.

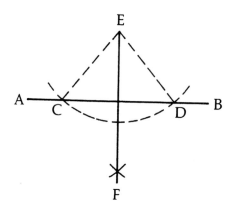

3. Bisect line AB. With A and B as centers, swing arcs intersecting at C and D. Draw line CD, which bisects line AB at point E.

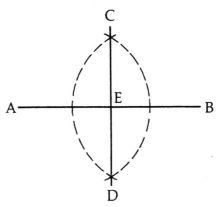

4. Bisect any angle. From C as center, swing arc AB, cutting the two sides of the angle. From A and B as centers, swing short arcs intersecting at D. Draw line CD, which bisects the angle.

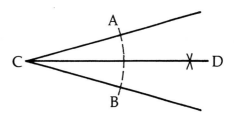

5. Construct a perpendicular to line AB from a given point C. With C as center, swing arc intersecting line AB at E and F. With E and F as centers, swing short arcs intersecting at D. Draw a line CD, which is perpendicular to line AB.

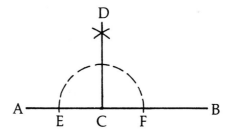

6. Find the center of a circle or arc. Draw any two chords, AB and CD. From A, B, C, and D as centers, swing arcs intersecting at E, F, and G, H. Draw lines through E, F, and G, H long enough to intersect at I, which is the center of circle or arc.

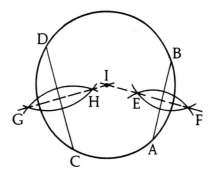

Surveying Techniques

Bearings of Property Lines

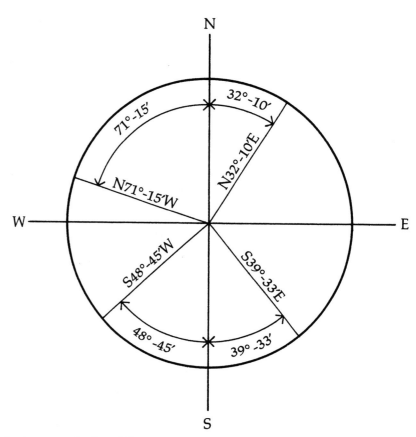

The bearing of a line falls in a quadrant: NE, SE, NW or SW; and is written, for example, N32°-10′E, or S48°-45′W. The line AB, below, relative to the compass, is in the NE quadrant, and is written Nx°-y′E.

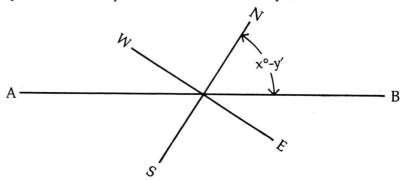

Converting Bearings to Azimuths

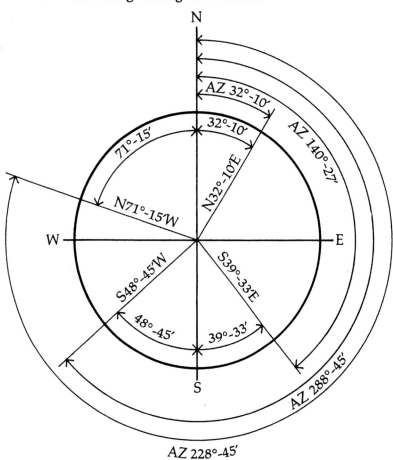

1. Bearing in NE quadrant—bearing and azimuth are the same.
2. Bearing in SE quadrant—subtract bearing reading from 180°.
3. Bearing in SW quadrant—add bearing reading to 180°.
4. Bearing in NW quadrant—subtract bearing from 360°.

Compute Interior Angles of Property Lines

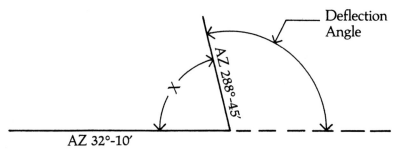

1. Subtract the azimuth of adjacent property lines and find the deflection angle.
2. If the deflection angle is less than 180°, subtract from 180° to find the interior angle "x".
3. If the deflection angle is greater than 180°, subtract 180° from the deflection angle to find "x".

Example: deflection angle = (288°-45′)−(32°-10′) = 256°-35′. Since the deflection angle is greater than 180°, use rule
(3): x = (256°-35′)−(180°) = 76°-35′.

Laying out of curves

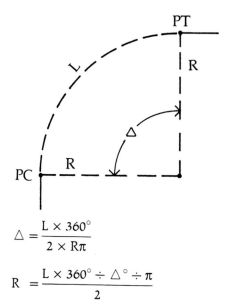

PC = point of curve (start)
PT = point of tangent
R = radius
△ = angle of intersection
L = arc (length of curve)

$$\triangle = \frac{L \times 360°}{2 \times R\pi}$$

$$R = \frac{L \times 360° \div \triangle° \div \pi}{2}$$

$$L = \frac{\triangle}{360°} \times 2R\pi$$

Trigonometric Properties

Plane trigonometry deals primarily with right triangles, making possible many more relationships than does the Pythagorean Theorum which was mentioned earlier in this section. Special formulas for the solution of oblique (non-right angle) triangles are given in Appendix C. The key to solving trigonometric problems is the following list of functions relating to the typical right triangle as illustrated.

$$\text{sine } A = \frac{\text{opposite}}{\text{hypotenuse}} = \frac{a}{c}$$

$$\text{cosine } A = \frac{\text{adjacent}}{\text{hypotenuse}} = \frac{b}{c}$$

$$\text{tangent } A = \frac{\text{opposite}}{\text{adjacent}} = \frac{a}{b}$$

$$\text{cotangent } A = \frac{\text{adjacent}}{\text{opposite}} = \frac{b}{a}$$

$$\text{secant } A = \frac{\text{hypotenuse}}{\text{adjacent}} = \frac{c}{b}$$

$$\text{cosecant } A = \frac{\text{hypotenuse}}{\text{opposite}} = \frac{c}{a}$$

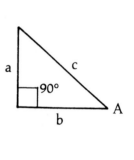

By using these relationships and the following trigonometric table, unknown angles and sides of right triangles may be easily found.

TABLE OF NATURAL TRIGONOMETRIC FUNCTIONS

Degrees	Sin	Cos	Tan	Cot	Sec	Csc	
0° 00′	.0000	1.0000	.0000	—	1.000	—	90° 00′
10	029	000	029	343.8	000	343.8	50
20	058	000	058	171.9	000	171.9	40
30	.0087	1.0000	.0087	114.6	1.000	114.6	30
40	116	9999	116	85.94	000	85.95	20
50	145	999	145	68.75	000	68.76	10
1° 00′	.0175	.9998	.0175	57.29	1.000	57.30	89° 00′
10	204	998	204	49.10	000	49.11	50
20	233	997	233	42.96	000	42.98	40
30	.0262	.9997	.0262	38.19	1.000	38.20	30
40	291	996	291	34.37	000	34.38	20
50	320	995	320	31.24	001	31.26	10
2° 00′	.0349	.9994	.0349	28.64	1.001	28.65	88° 00′
10	378	993	378	26.43	001	26.45	50
20	407	992	407	24.54	001	24.56	40
30	.0436	.9990	.0437	22.90	1.001	22.93	30
40	465	989	466	21.47	001	21.49	20
50	494	988	495	20.21	001	20.23	10
3° 00′	.0523	.9986	.0524	19.08	1.001	19.11	87° 00′
10	552	985	553	18.07	002	18.10	50
20	581	983	582	17.17	002	17.20	40
30	.0610	.9981	.0612	16.35	1.002	16.38	30
40	640	980	641	15.60	002	15.64	20
50	669	978	670	14.92	002	14.96	10
4° 00′	.0698	.9976	.0699	14.30	1.002	14.34	86° 00′
10	727	974	729	13.73	003	13.76	50
20	756	971	758	13.20	003	13.23	40
30	.0785	.9969	.0787	12.71	1.003	12.75	30
40	814	967	816	12.25	003	12.29	20
50	843	964	846	11.83	004	11.87	10
5° 00′	.0872	.9962	.0875	11.43	1.004	11.47	85° 00′
10	901	959	904	11.06	004	11.10	50
20	929	957	934	10.71	004	10.76	40
30	.0958	.9954	.0963	10.39	1.005	10.43	30
40	987	951	992	10.08	005	10.13	20
50	.1016	948	.1022	9.788	005	9.839	10
6° 00′	.1045	.9945	.1051	9.514	1.006	9.567	84° 00′
10	074	942	080	9.255	006	9.309	50
20	103	939	110	9.010	006	9.065	40
30	.1132	.9936	.1139	8.777	1.006	8.834	30
40	161	932	169	8.556	007	8.614	20
50	190	929	198	8.345	007	8.405	10
7° 00′	.1219	.9925	.1228	8.144	1.008	8.206	83° 00′
10	248	922	257	7.953	008	8.016	50
20	276	918	287	7.770	008	7.834	40
30	.1305	.9914	1317	7.596	1.009	7.661	30
40	334	911	346	7.429	009	7.496	20
50	363	907	376	7.269	009	7.337	10
8° 00′	.1392	.9903	.1405	7.115	1.010	7.185	82° 00′
10	421	899	435	6.968	010	7.040	50
20	449	894	465	6.827	011	6.900	40
30	.1478	.9890	.1495	6.691	1.011	6.765	30
40	507	886	524	6.561	012	6.636	20
50	536	881	554	6.435	012	6.512	10
9° 00′	.1564	.9877	.1584	6.314	1.012	6.392	81° 00′
	Cos	Sin	Cot	Tan	Csc	Sec	Degrees

Degrees	Sin	Cos	Tan	Cot	Sec	Csc	
36° 00′	.5878	.8090	.7265	1.376	1.236	1.701	54° 00′
10	901	073	310	368	239	695	50
20	925	056	355	360	241	688	40
30	.5948	.8039	.7400	1.351	1.244	1.681	30
40	972	021	445	343	247	675	20
50	995	004	490	335	249	668	10
37° 00′	.6018	.7986	.7536	1.327	1.252	1.662	53° 00′
10	041	969	581	319	255	655	50
20	065	951	627	311	258	649	40
30	.6088	.7934	.7673	1.303	1.260	1.643	30
40	111	916	720	295	263	636	20
50	134	898	766	288	266	630	10
28° 00′	.6157	.7880	.7813	1.280	1.269	1.624	52° 00′
10	180	862	860	272	272	618	50
20	202	844	907	265	275	612	40
30	.6225	.7826	.7954	1.257	1.278	1.606	30
40	248	808	.8002	250	281	601	20
50	271	790	050	242	284	595	10
39° 00′	.6293	.7771	.8098	1.235	1.287	1.589	51° 00′
10	316	753	146	228	290	583	50
20	338	735	195	220	293	578	40
30	.6361	.7716	.8243	1.213	1.296	1.572	30
40	383	698	292	206	299	567	20
50	406	679	342	199	302	561	10
40° 00′	.6428	.7660	.8391	1.192	1.305	1.556	50° 00′
10	450	642	441	185	309	550	50
20	472	623	491	178	312	545	40
30	.6494	.7604	.8541	1.171	1.315	1.540	30
40	517	585	591	164	318	535	20
50	539	566	642	157	322	529	10
41° 00′	.6561	.7547	.8693	1.150	1.325	1.524	49° 00′
10	583	528	744	144	328	519	50
20	604	509	796	137	332	514	40
30	.6626	.7490	.8847	1.130	1.335	1.509	30
40	648	470	899	124	339	504	20
50	670	451	952	117	342	499	10
42° 00′	.6691	.7431	.9004	1.111	1.346	1.494	48° 00′
10	713	412	057	104	349	490	50
20	734	392	110	098	353	485	40
30	.6756	.7373	.9163	1.091	1.356	1.480	30
40	777	353	217	085	360	476	20
50	799	333	271	079	364	471	10
43° 00′	.6820	.7314	.9325	1.072	1.367	1.466	47° 00′
10	841	294	380	066	371	462	50
20	862	274	435	060	375	457	40
30	.6884	.7254	.9490	1.054	1.379	1.453	30
40	905	234	545	048	382	448	20
50	926	214	601	042	386	444	10
44° 00′	.6947	.7193	.9657	1.036	1.390	1.440	46° 00′
10	967	173	713	030	394	435	50
20	988	153	770	024	398	431	40
30	.7009	.7133	.9827	1.018	1.402	1.427	30
40	030	112	884	012	406	423	20
50	050	092	942	006	410	418	10
45° 00′	.7071	.7071	1.0000	1.000	1.414	1.414	45° 00′
	Cos	Sin	Cot	Tan	Csc	Sec	Degrees

Degrees	Sin	Cos	Tan	Cot	Sec	Csc	
27° 00′	.4540	.8910	.5095	1.963	1.122	2.203	63° 00′
10	566	897	132	949	124	190	50
20	592	884	169	935	126	178	40
30	4617	.8870	.5206	1.921	1.127	2.166	30
40	643	857	243	907	129	154	20
50	669	843	280	894	131	142	10
28° 00′	.4695	.8829	.5317	1.881	1.133	2.130	62° 00′
10	720	816	354	868	143	118	50
20	746	802	392	855	136	107	40
30	.4772	.8788	.5430	1.842	1.138	2.096	30
40	797	774	467	829	140	085	20
50	823	760	505	816	142	074	10
29° 00′	.4848	.8746	.5543	1.804	1.143	2.063	61° 00′
10	874	732	581	792	145	052	50
20	899	718	619	780	147	041	40
30	.4924	.8704	.5658	1.767	1.149	2.031	30
40	950	689	696	756	151	020	20
50	975	675	735	744	153	010	10
30° 00′	.500	.8660	.5774	1.732	1.155	2.000	60° 00′
10	025	646	812	720	157	1.990	50
20	050	631	851	709	159	980	40
30	.5075	.8616	.5890	1.698	1.161	1.970	30
40	100	601	930	686	163	961	20
50	125	587	969	675	165	951	10
31° 00′	.5150	.8572	.6009	1.664	1.167	1.942	59° 00′
10	175	557	048	653	169	932	50
20	200	542	088	643	171	923	40
30	.5225	.8526	.6128	1.632	1.173	1.914	30
40	250	511	168	621	175	905	20
50	275	496	208	611	177	896	10
32° 00′	.5299	.8480	.6249	1.600	1.179	1.887	58° 00′
10	324	465	289	590	181	878	50
20	348	450	330	580	184	870	40
30	.5373	.8434	.6371	1.570	1.186	1.861	30
40	398	418	412	560	188	853	20
50	422	403	453	550	190	844	10
33° 00′	.5446	.8387	.6494	1.540	1.192	1.836	57° 00′
10	471	371	536	530	195	828	50
20	495	355	577	530	197	820	40
30	.5519	.8339	.6619	1.511	1.199	1.812	30
40	544	323	661	501	202	804	20
50	568	307	703	1.492	204	796	10
34° 00′	.5592	.8290	.6745	1.483	1.206	1.788	56° 00′
10	616	274	787	473	209	781	50
20	640	258	830	464	211	773	40
30	.5664	.8241	.6873	1.455	1.213	1.766	30
40	688	225	916	446	216	758	20
50	712	208	959	437	218	751	10
35° 00′	.5736	.8192	.7002	1.428	1.221	1.743	55° 00′
10	760	175	046	419	223	736	50
20	783	158	089	411	226	729	40
30	.5807	.8141	.7133	1.402	1.228	1.722	30
40	831	124	177	393	231	715	20
50	854	107	221	385	233	708	10
36° 00′	.5878	.8090	.7265	1.376	1.236	1.701	54° 00′
	Cos	Sin	Cot	Tan	Csc	Sec	Degrees

TABLE OF NATURAL TRIGONOMETRIC FUNCTIONS (*Continued*)

Degrees	Sin	Cos	Tan	Cot	Sec	Csc	
18° 00′	.3090	.9511	.3249	3.078	1.051	3.236	72° 00′
10	118	502	281	047	052	207	50
20	145	492	314	018	053	179	40
30	.3173	.9843	.3346	2.989	1.054	3.152	30
40	201	474	378	960	056	124	20
50	228	465	411	932	057	098	10
19° 00′	.3256	.9455	.3443	2.904	1.058	3.072	71° 00′
10	283	446	476	877	059	046	50
20	311	436	508	850	060	021	40
30	.3388	.9426	.3541	2.824	1.061	2.996	30
40	365	417	574	798	062	971	20
50	393	407	607	773	063	947	10
20° 00′	.3420	.9397	.3640	2.747	1.064	2.924	70° 00′
10	448	387	673	723	065	901	50
20	475	377	706	699	066	878	40
30	.3502	.9367	.3739	2.675	1-068	2.855	30
40	529	356	772	651	069	833	20
50	557	346	805	628	070	812	10
21° 00′	.3584	.9336	.3839	2.605	1.071	2.790	69° 00′
10	611	325	872	583	072	769	50
20	638	315	906	560	074	749	40
30	.3665	.9304	.3939	2.539	1.075	2.729	30
40	692	293	973	517	076	709	20
50	719	283	.4006	496	077	689	10
22° 00′	.3746	.0272	.4040	2.475	1.079	2.669	68° 00′
10	773	261	074	455	080	650	50
20	800	250	108	434	081	632	40
30	.3827	.9239	.4142	2.414	1.082	2.613	30
40	854	228	176	394	084	595	20
50	881	216	210	375	085	577	10
23° 00′	.3907	.9205	.4245	2.356	1.086	2.559	67° 00′
10	934	194	279	337	088	542	50
20	961	182	314	318	089	525	40
30	.3987	.9171	.4348	2.300	1.090	2.508	30
40	.4014	159	383	282	092	491	20
50	041	147	417	264	093	475	10
24° 00′	.4067	.9135	.4452	2.246	1.095	2.459	66° 00′
10	094	124	487	229	096	443	50
20	120	112	522	211	097	427	40
30	.4147	.9100	.4557	2.194	1.099	2.411	30
40	173	088	592	177	100	396	20
50	200	075	628	161	102	381	10
25° 00′	.4226	.9063	.4663	2.145	1.103	2.366	65° 00′
10	253	051	699	128	105	352	50
20	279	038	734	112	106	337	40
30	.4305	.9026	.4770	2.097	1.108	2.323	30
40	331	013	806	081	109	309	20
50	358	.001	841	066	111	295	10
26° 00′	.4384	.8988	.4877	2.050	1.113	2.281	64° 00′
10	410	975	913	035	114	268	50
20	436	962	950	020	116	254	40
30	.4462	.8949	.4986	2.006	1.117	2.241	30
40	488	936	.5022	1.991	119	228	20
50	514	923	059	977	121	215	10
27° 00′	.4540	.8910	.5095	1.963	1.122	2.203	63° 00′
	Cos	Sin	Cot	Tan	Csc	Sec	Degrees

Degrees	Sin	Cos	Tan	Cot	Sec	Csc	
9° 00′	.1564	.9877	.1584	6.314	1.012	6.392	81° 00′
10	593	872	614	197	013	277	50
20	622	868	644	084	013	166	40
30	.1650	.9863	.1673	5.976	1.014	6.059	30
40	679	858	703	871	014	5.955	20
50	708	853	733	769	015	855	10
10° 00′	.1736	.9848	.1763	5.671	1.015	5.759	80° 00′
10	765	843	793	576	016	665	50
20	794	838	823	485	016	575	40
30	.1822	.9833	.1853	5.396	1.017	5.487	30
40	851	827	883	309	018	403	20
50	880	822	914	226	018	320	10
11° 00′	.1908	.9816	.1944	5.145	1.019	5.241	79° 00′
10	937	811	974	066	019	164	50
20	965	805	.2004	4.989	020	089	40
30	.1994	.9799	.2035	4.915	1.020	5.016	30
40	.2022	793	065	843	021	4.945	20
50	051	787	095	773	022	876	10
12° 00′	.2079	.9781	.2126	4.705	1.022	4.810	78° 00′
10	108	775	156	638	023	745	50
20	136	769	186	574	024	682	40
30	.2164	.9763	.2217	4.511	1.024	4.620	30
40	193	757	247	449	025	560	20
50	221	750	278	390	026	502	10
13° 00′	.2250	.9744	.2309	4.331	1.026	4.445	77° 00′
10	278	737	339	275	027	390	50
20	306	730	370	219	028	336	40
30	.2334	.9724	.2401	4.165	1.028	4.284	30
40	363	717	432	113	029	232	20
50	391	710	462	061	030	182	10
14° 00′	.2419	.9703	.2493	4.011	1.031	4.134	76° 00′
10	447	696	524	3.962	031	086	50
20	476	689	555	914	032	039	40
30	.2504	.9681	.2586	3.868	1.033	3.994	30
40	532	674	617	821	034	950	20
50	560	667	648	776	034	906	10
15° 00′	.2588	.9659	.2679	3.732	2.732	3.864	75° 00′
10	616	652	711	689	036	822	50
20	644	644	742	647	037	782	40
30	.2672	.9636	.2773	3.606	1.038	3.742	30
40	700	628	805	566	039	703	20
50	728	621	836	526	039	665	10
16° 00′	.2756	.9613	.2867	3.487	1.040	3.628	74° 00′
10	784	605	899	450	041	592	50
20	812	596	931	412	042	556	40
30	.2840	.9588	.2962	3.376	1.043	3.521	30
40	868	580	994	340	044	487	20
50	896	572	.3026	305	045	453	10
17° 00′	.2924	.9563	.3057	3.271	1.046	3.420	73° 00′
10	952	555	089	237	047	388	50
20	979	546	121	204	048	356	40
30	.3007	.9537	.3153	3.172	1.049	3.326	30
40	035	528	185	140	049	295	20
50	062	520	217	108	050	265	10
18° 00′	.3090	.9511	.3249	3.078	1.051	3.236	72° 00′
	Cos	Sin	Cot	Tan	Csc	Sec	Degrees

INDEX